Uniqueness
and Diversity
in Human
Evolution

Spectral tarsier.

Charles E. Oxnard

Uniqueness and Diversity in Human Evolution:

Morpho-metric Studies of Australo-pithecines

The University of Chicago Press
Chicago London

CHARLES OXNARD is dean of the College and professor in the Department of Anatomy, the Department of Anthropology, and the Committee on Evolutionary Biology at the University of Chicago. He has published numerous articles and is the author of *Form and Pattern in Human Evolution* (also published by the University of Chicago Press).

The University of Chicago Press. Chicago 60637

The University of Chicago Press, Ltd., London

© 1975 by The University of Chicago

All rights reserved. Published 1975

Printed in the United States of America

International Standard Book Number: 0-226-64253-4

Library of Congress Catalog Card Number: 74-16689

Library of Congress Cataloging in Publication Data

Oxnard, Charles.
    Uniqueness and diversity in human evolution.

    Bibliography: p.
    Includes index.
    1. Australopithecines. I. Title. [DNLM: 1. Evo-
lution. 2. Paleontology.   QE740 098u]
GN283.096   573.2   74-16689
ISBN 0-226-64253-4

Contents

Preface

This book is in part a sequel to a previous study of mine: *Form and Pattern in Human Evolution: Some Mathematical, Physical and Engineering Approaches*. That volume comprised a series of descriptions of modern methods available for the study of form and function and applied them to the investigation of human evolution through the examination of bones and fossils. It was specifically written so that the methods could be understood without mathematical, physical, or engineering expertise. At the same time as it demonstrated the nature of the various techniques, however, it also attempted to show how they may be used in studies of the evolution of function in primates and man, and it gave some indication of the nature of the new information that may flow from their use. In the hope of stimulating interest in yet newer approaches, the book purposely described some methods that are not commonly used in evolutionary biology.

The present book follows as a description of a series of studies in which the most prominent, and the most fully understood, of the methods—the multivariate statistical approach—has been applied. The method is used not only to provide background information about man and other living primates but also about a group of African fossils: the australopithecines. This book, then, describes the results of a number of morphometric studies and attempts to place the australopithecines in some relationship to modern primates from the viewpoint of the evolution of locomotor function.

As with the previous book, the results must be considered as tentative, if only because the numbers of anatomical fragments that have been analyzed by these methods are still small compared to the total number of finds known. In addition the results suffer from the deficit that in some cases the analyses have had to be based upon data obtained by other individuals. This latter criticism is, however, also a strength in that it indicates independence of analysis from data collection.

Bearing these caveats in mind, the reader will note that the results are controversial. Whereas the conventional wisdom about human evolution depends upon the (apparent) marked similarity between modern man and the various australopithecine fossils, the studies here indicate that these fossils are uniquely different from modern man in many major respects. Again the conventional wisdom has it that when the australopithecine fossils do depart from the condition in modern man, then they do so in the direction of the African apes, our nearest genetic relatives. The present study indicates that this may not be the case; more significant resemblances to and differences from other primates can be discovered through the multivariate statistical approach. We may well ask: What is the biological meaning of this new set of morphological relationships?

This book naturally follows from the first. But it also depends in large part upon an invitation from Steven J. Gould for my participation in a symposium, *Evolutionary Development of Form and Symmetry*, at the First International Congress of Systematic and Evolutionary Biology in Boulder, Colorado, 1973. Preparations for that symposium and the stimulus which followed from the confluence of workers whom Professor Gould had gathered together at that time, played a major role. The basic format of the book has also been influenced by the further development of my graduate course on the analysis of biological form and pattern at the University of Chicago. As such it owes much to those students who have taken the course over the last few years and especially to my own graduate students, who have provided much discussion and criticism. Perhaps most of all, however, this second book, along with the more recent developments of my investigations, has been stimulated by my colleagues in Evolutionary Morphology in the Department of Anatomy, the University of Chicago; Professors Ronald Singer, Jack T. Stern, Jr., and R. Eric Lombard have participated in joint courses, have exchanged views about research strategies and tactics, and have discussed and criticized many manuscripts. In particular they have read several drafts of this work and are responsible for much improvement.

The broader background of research on which this book is grounded owes much to my collaborators in England. Thus Professor Eric H. Ashton, Dr. R. M. Flinn, and Mr. T. F. Spence at the University of Birmingham, and Professor Lord Zuckerman, Secretary of the Zoological Society of London and at the University of East Anglia continue to be collaborators in some of my research. I am also happy to acknowledge the collaboration that has developed with Professor F. Peter Lisowski at the University of Hong Kong. These joint investigations have all contributed to this book.

In addition to the collaboration mentioned above, it is especially appropriate to acknowledge Professor Lord Zuckerman's stimulus to and continuing interest in all of my work, not only during the period when he was Sands Cox Professor of Anatomy at the University of Birmingham, but also during my studies in the United States.

It is also an especial pleasure to acknowledge the help and expertise of my personal research assistant, Miss Joan Hives, who has taken part in my investigations for many years and who has contributed greatly in the preparation of this book. She has willingly undertaken the many technical, artistic, and bibliographic tasks that have arisen, and has contributed largely to the original studies upon which the book is based.

Many others have contributed through criticism and 'technical assistance. Dr. Gene Albrecht must especially be thanked for writing and running a number of the computer programs necessary for the analyses. The photographic expertise of Miss Shirley Aumiller and Mr. Melvin Oster is gratefully acknowledged. The biological sciences staff of the Joseph R. Regenstein Library at the University of Chicago has also been most helpful.

This work has been supported by the United States Public Health Service Grants HD 0054 (1960–1966), HD 02852 (1968–1971) and AM 16805 (1974–1976), by the National Science Foundation Grant GS 30508 (1971–1976), and by grants from the Louis Block Bequest Fund (1966–1967) and the Dr. W. C. and C. A. Abbott Memorial Fund (1966–1967) of the University of Chicago. A travel grant partly related to collaborative research was received from the Wenner-Gren Foundation for Anthropological Research (1970).

1   Relationships between Animal Form and Function

The Direct Approach in Understanding Animal Form

One of the major problems facing modern organismal biology is understanding the relationships that exist between the structures that organisms possess and the functions that they display. This problem may be studied at a number of different levels. Implicit within it is the understanding of the mechanisms that lead, within a given ontogeny, to a particular functional-structural relationship. Once purely descriptive of the changes that occur during development, present investigations in this area encompass an understanding of the underlying developmental processes that give rise to organismic form and pattern. At another grade, and, of course, related to this, is the study of how the functional-structural bond changes over geological time. Again, though once principally aimed at documenting the course of evolution, these studies are today often aimed at investigating mechanism and process within evolution. Yet a third mode of research helps us understand the functional-structural continuum—that is, the direct impact upon structure of function through biomechanics; this has relatively recently received an impetus from the discovery of the various mechano-electric phenomena within biological materials. None of the above attacks upon the problems of macrobiology is isolated; each depends upon the others for full understanding. Yet the disentangling of functional-structural interrelationships often starts with the understanding that animal mechanics supplies one set of constraints within which ontogeny and evolution must operate.

Biomechanical aspects of form can be studied at relatively simple levels. For instance, physical principles relating to size, load, and properties of materials provide constraints that determine some of the limits of biological form. We understand, thus, that land animals the size of the biggest of blue whales, if constructed of conventional biological materials, cannot exist; we accept, likewise, that there are constraints, based upon physical laws, which bound the order of magnitude of insects as we know them on earth.

But many biomechanical studies have become considerably more complicated as a result of the capability for study that is provided by modern developments in such fields as, for example, equipment miniaturization and computer simulation. Thus by using more complex methods for modeling, in the computer, the physical parameters of muscles and joints during movement, Stern (personal communication) suggests that the age-old idea of equally proportioned but differently sized animals having identical maximum speeds (demanded by simple considerations of dimension, mass and velocity) may not be true; he suggests that there may be a 10 percent disagreement that can be based upon good biological concepts. Actual determinations of animal speeds during locomotion are not yet good enough to test the reality of this theory.

Experimental methods such as cineradiography are also today providing information about biomechanisms that hone our understanding and in many cases provide entirely new views of traditional subjects. Thus studies of chimpanzee bipedalism (Jenkins 1972) using such methods have defined the difference that exists between the way some apes walk when they attempt upright movement and the bipedal gait of modern man. Thus Jenkins has shown us that bipedal chimpanzees do not adopt a new method of walking when they move bipedally; rather do they retain the pattern of femoral flexion and extension that is characteristic of their own quadrupedalism and apply it within a posture that includes a reoriented pelvis. Such differences allow us to define in detail how the biomechanical elements of the structure of the human hip differ from that of an ape.

Yet further methods, such as experimental stress analyses utilizing strain-gages in vivo (Lanyon 1971, 1972, and 1973), are demonstrating that not only does the vertebral column help to bear obvious loads associated with posture and locomotion, but also that each individual vertebral body expands and contracts like a concertina as a result of apparently small cyclical forces, hitherto generally neglected by researchers, such as those of the heart beat and respiratory rhythm.

The combination of a number of such experimental methods can be most fruitful. One excellent example here is the studies of Karel Liem (1970, 1973) who, using the interplay of electromyography, cinematography, and surgical interference, has demonstrated how minor differences in mechanical units of the jaws of nandid fishes provide a chain reaction of functionally co-adaptive alterations in neighboring and remote anatomical areas that is the raw material for the apparently explosive radiation of the leaf fishes within various African Lakes. What appear to be major differences between leaf fishes as varied as those that swallow large prey in toto and those that eat the scales of living prey, can be shown to be, at least in part, a resultant of rather minor differences in rate of growth or degree of ossification of small cranial skeletal elements. Many other studies could be cited that today are similarly providing information about functional-structural interrelationships in living species.

The Direct Approach in the Study of Fossils

When, however, our interests stray towards making functional-structural assessments of fossil forms, a number of marked constraints enter into methodology. The direct approaches just discussed are capable of providing information about fossils only to the extent that analogy with the extant forms can be utilized. Though, for instance, we have no known extant biological equivalent to the "sails" of the pelycosaurs that can be used experimentally to investigate their possible function, we may make, fortunately, in the case of many of the problems relating to primate evolution, reasonable inferences from models provided by study of related living species.

Thus to expand the example quoted above, Jenkins (1972) has indicated that, in addition to the difference he has perceived between bipedal walking in the chimpanzee and modern man, pertinent structural elements of the australopithecines[1] may be indicative of a type of bipedalism that differs markedly from that of man. Jenkins's evidence can only be based indirectly, of course, upon the known structures together with reasoned guesses at the functional analogies.

1. Such as the orientation of the margin of the femoral head relative to the acetabular rim suggesting that the fossil may have employed relatively more abducted femoral excursion than occurs in man, and which exists in the chimpanzee.

Direct approaches can also be made by modeling biomechanical situations in the fossils themselves. Such models can be entirely theoretical. Thus utilizing theoretical stress analysis, Preuschoft (1972) has shown, for instance, that the form of the metacarpals in some dryopithecine apes is well adapted to use in palmigrade postures and gaits. The adoption of digitigrade modes of locomotion as in terrestrial monkeys today, or of knuckle-walking as in extant African apes, would appear to lead to greater stresses than are efficient in restricted parts of the fossil metacarpals. Does this indicate the unlikelihood of digitigrade or knuckle-walking locomotor patterns in dryopithecine apes? More positively, does it confirm palmigrady in these creatures? Are we justified in making the appropriate inferences?

Of course such an analysis, based as it is upon theoretical concepts of metacarpals approximating regular solid cylinders and depending upon educated guesses about the forces existing within tendons, cannot give precise information. But we must live with these approximations; the hope is that the answer provided by the theoretical ideas is not too far from the truth.

In a similar way, the methods of experimental stress analysis can also be applied to fossil shapes in comparisons with the structures of extant species the functions of which are known. Thus Oxnard (1973a) has shown that the complicated architecture of the finger bones of higher primates relates fairly well to primary functions during locomotion. The finger bones of the chimpanzee and gorilla are efficient within static knuckle-walking simulations, while those of the orangutan are not. These same architectures are, for the chimpanzee and gorilla, somewhat inefficient in a static hanging-climbing context, whereas those of the orangutan are here efficient. Man is inefficient in both situations; of course we know that he practices neither. A finger bone of the Olduvai hand (*"Homo habilis"*—an australopithecine) is inefficient at knuckle-walking but efficient in the hanging-climbing mode. Again, does this suggest, negatively, that this species is unlikely to be a knuckle-walker? Does this suggest, positively, that, whatever else these hands were doing, they had not yet lost abilities for climbing? Again, may we make the appropriate inferences?

Like the theoretical stress investigations, this experimental study also is subject to qualifications; here the bones do not have to be regarded as simple cylinders, greater reality is achieved in terms of the shapes of the elements. But in this case, the biomechanical simulation is somewhat poorer, fewer forces can be considered, they are viewed again in a static situation and, as before, their magnitude can only be guessed at. But this particular set of studies does introduce the use of serial increase in reality of the experimental simulations in an attempt to "control" for the effects of approximations (Oxnard 1973a).

Both types of direct study suffer from the major deficiency that the regions modeled, whether theoretically or by analogy, are treated as though they were uniform elastic bodies. Bone is not, of course, a homogeneous, isotropic material operating as a uniform elastic body under "infinite beam theory." There is now every reason for believing that bone may be better considered as an anisotropic and poroelastic material (Nowinski and Davis 1970) and that for theoretical stress analysis

studies "finite element elastic theory" may be more appropriate, especially for com-
plexly shaped regions (Rybicki, Simonen, and Weis 1972; Farah, Craig, and
Sikarskie 1973). Again, however, we hope that the inferences derived from models,
whether theoretical or experimental, will not be rendered too inaccurate by the
simplifications adopted.

Inductive Methods in Understanding Animal Form

A second, indirect, approach is also available for the study of animal form. This, the
inductive method, consists of allowing the structures to speak, as it were, for themselves. This approach is a good deal less directly aimed at biomechanics than the
experimental methods discussed above. And because the structures do not always
speak very clearly when they tell us about themselves, we may run into different sets
of problems. However, although the results must be inferred from complex comparisons, rather than neatly displayed by experiment, the inductive approach has some
major facets in which it is decidedly superior to the experimental. Thus it is more
easily able to deal with populations rather than individuals, something that is usually
prohibitive for more than a small number of cases in experimental work. It is better
able to deal with a wide diversity of anatomical structures and taxonomic groups
such as can scarcely be included in carefully planned experimental studies with
adequate controls. It is capable of dealing with fragmentary and incomplete specimens in a manner difficult to arrange in experimental studies. For these reasons,
therefore, it is especially appropriate for the study of fossils. And it is to this general
approach that attention is directed in this book.

However, we must acknowledge that the inductive method is not at variance with
the direct techniques. In fact, it is clear that the converse applies: these various approaches are truly complementary rather than competitive; concordance among
them is an important strength in evaluating results; disagreements should make each
look to the beam in his own eye.

The inductive approach may involve a methodology as simple as associating,
visually, functional observations with morphological information. For example,
animals with curved but non-buttressed phalanges are usually capable climbers
(gibbon, spider monkey, orangutan); ergo, fossils with curved non-buttressed
phalanges may have been capable climbers. This is a well-known methodology that
has been practiced for decades. It may run into difficulties, however, once biological
problems become more refined, as indeed has this particular example: chimpanzees
and gorillas also have curved phalanges, but in these cases with special buttresses
that seem to be related to knuckle-walking. Where does the consideration of buttresses start and end in the examination of fossil forms in which the incipient natures
of the morphological differences and beginning behaviors may be the precise information that is needed in an evolutionary context? And in any case, although a
rough association between function and morphology readily springs to mind whatever the function and whatever the morphology happen to be, once we become
interested in shades of differences, the complex nature of bone form and pattern,

and the complicated series of developmental, evolutionary and biomechanical processes that are responsible for its precise expression, defy the facile explanation. Under such circumstances a number of other techniques for characterizing bone form and pattern may also be useful.

Some of these methods have been discussed by Oxnard (1972a, b; 1973a, c) and they include a series of techniques for digitizing complex shapes and then asking ad hoc questions about them for comparative purposes (e.g., see Shelman and Hodges 1970). There are also methods of using specific mathematical transformations to aid in the process of characterizing and comparing complex shapes and structures. One is the medial axis transformation, in which a shape may be characterized by the "function" (mathematical) of its "skeleton" (the structure resulting from collapsing the shape into itself in a prescribed manner). Another method is the well-known use of distorted cartesian coordinate diagrams, as displayed by D'Arcy Thompson (1917, 1942) and later adapted in an interesting algebraic manner by Sneath (1967) for comparing statistical "best fits" of one shape with another. Still other workers have utilized Fourier transformations; Lestrel (1973) attempts this in a one-dimensional manner for analyzing skull outlines; it had also already been used in its two-dimensional form for the rather different problem of characterizing complex cancellous internal structure of vertebrae (Oxnard, 1970, 1972b, work in progress). These last studies have allowed the recognition of the patterns of trabeculae in individual vertebrae with a sensitivity that can visualize differences between vertebrae as similar as the second and fourth lumbar in man. It seems as though certain structural features of these complex cancellous patterns in the second lumbar vertebrae can rather readily be related to bending in this vertebra during normal posture and movement; in the fourth lumbar vertebra the absence of such elements may be associated with a relative resultant lack of bending owing to the closer proximity of this vertebra to the average line of gravity of the body.

One particular method, the multivariate statistical approach, has been utilized fairly extensively for the characterization of bone form. And although this method has been used by a number of workers in different anatomical regions purely as a method of characterizing and comparing shapes (multivariate morphometrics), it has been used in a rather special manner in our laboratories for examining postcranial skeletal elements; in these particular studies the features used for comparisons are chosen in such a way as to maximize the functional content of the data.

This can be attempted by obtaining such information as may exist about the locomotor behaviors of living forms and then using this to make some assessment of the biomechanics of the different anatomical regions under study. The biomechanical deductions can be allied to morphological studies of muscles and bones in order to choose bony dimensions that may reflect biomechanics more than other aspects of bone biology. Multivariate study of these dimensions in both living and fossil species may then allow us to make appropriate biomechanical assessments that may provide inductive information about posture and locomotion in the fossils. A necessary first step in this type of study is some understanding of the very complex relationship between structure and behavior.

The Structure-Behavior Equation

There is no doubt that the locomotor behavior of an animal is, on a gross level, con-trolled by the anatomy of the animal.[2] A human with phocomelia of the upper ex-tremities is characterized by a much more restricted locomotor repertoire than a normal individual, but this is a very dramatic and unusual example. If upper limb phocomelia were adaptive, then one might anticipate a substantial change over many generations in the way the lower limbs (and other regions of the body) are used in movement and rest. In reality, however, there are no well-documented instances of morphological changes leading to new patterns of locomotor behavior. On the whole, as pointed out by Mayr (1963), it seems correct to state "that behavior movements often precede phylogenetically the special structures which make these movements particularly conspicuous."

An example of this phenomenon in the primates is the prehensile tail. Certainly this structure allows a different approach to locomotion. Yet it was not the develop-ment of a prehensile tail which then led some monkey to curl its tail around a branch. The reality of the situation is that cebids without a specialized tail mor-phology engage in various behaviors involving wrapping of the tail around some object (a comrade's tail, its mother, or a branch). Indeed, even the Old World crab-eating monkey (*Macaca irus*) can use its tail to grasp small objects loosely and does use it (in captivity) for various manipulative activities. One can speculate con-fidently that it may have been these kinds of behaviors which preceded the mor-phological changes seen in the tails of the prehensile-tailed New World monkeys; it may have been the existence of these behaviors which were necessary in order for the morphological changes in the tail to have been selected.

Thus locomotor morphology is the result of an evolutionary selective process whose proximate end[3] is to improve the performance of some pre-existing be-havior. Locomotor morphology is also, however, the result of ontogenetic processes whose effect is continually to improve structural relationships for mechanical efficiency in the performance of pre-existing net behaviors. In its most simple form therefore we can express the relationship between morphology and behavior thus:

$$\text{Morphology} = \text{f(behavior)}$$
$$y = \text{f}(x)$$

An imaginary example might be the possession of longitudinally curved caudal vertebrae by animals that hang by their tails—longitudinally curved caudal vertebrae might be related to hanging in the same way that longitudinally curved phalanges are. It is clear however that the exact structure of an individual animal is not only achieved in response to behavioral requirements, but is also influenced by epigenetic, mutational and stochastic processes. We can thus modify the equation

$$\text{morphology} = \text{f(behavior)} + \text{error}$$
$$y = \text{f}(x) + e$$

2. Discussion with J. T. Stern, Jr., has materially improved this section (see also Stern and Oxnard 1973).
3. The ultimate end of selection is increased reproductivity.

where the error represents processes random with respect to behavior. An equivalent imaginary example might be the possession of curved caudal vertebrae with transverse foramina in animals that hang by their tails—transverse foramina might be merely an unusual phenomenon related to the serial nature of a vertebra. In examining this relationship we may try to minimize e by studying animals of close phylogenetic relationship (thus basically similar genetic substrate). We are encouraged that e is not unmanageably large by the numerous examples of parallelism and convergence which demonstrate that animals with different evolutionary histories can respond to similar selective pressures with the production of remarkably similar morphologies.

An equation such as the above does not necessarily imply a one-to-one relationship between morphology and behavior. First, of course, the error factor in the equation may lead to similarities in morphologies when the behaviors are distinct. But it is also clearly theoretically possible for two different behaviors to share the same ideal morphology (error disregarded). In other words, the equation relating morphology to behavior might be analogized to a quadratic equation

$$y = ax^2 + bx + e$$

in which a similar y (morphology), can result from two different positive real values of x (behavior). For instance curved phalanges (at a superficial level of description) are found both in animals that hang and those that knuckle-walk.[4] Higher order equations have proportionately more solutions. The reverse may presumably also occur, two different adaptive morphologies (error disregarded) being associated with identical behaviors.[5]

The nature of this relationship becomes even more complicated if we consider alterations in the *forms* of the morphologies, on the one hand, and the behaviors, on the other. Instead of representing these *forms* by a single y or x, we may think of them as more complex—say, as vectors, tensors, or matrices of y's and x's. For instance, earlier studies, essentially based upon natural history observations, have suggested that behaviors can be grouped in such a way that several different animals can be placed together in single clusters. This has been most frequently attempted with locomotion, and a series of workers have suggested locomotor groupings of the primates. Such might include a group of runners, for instance, or of leapers. In a similar way, but in far greater detail, earlier studies of morphology have suggested that animals can be grouped in relation to the structures that they share in common. This morphological grouping usually is rather easier to see than is the behavioral grouping just alluded to. And there is, of course, in the case of adaptive features of the postcranial skeleton, great temptation to read an association between these two sets of groupings.

4. It is only more detailed study of y (morphology) that shows that these two morphologies are not in fact identical (Tuttle 1969; Oxnard 1973a).
5. Again however, it is unlikely that this really occurs. For instance, at least two different morphologies have been shown to exist for vertical clinging and leaping prosimians. Examination of vertical clinging and leaping behavior suggests it may be done in distinctly different ways (Oxnard 1973b).

More recent studies by a number of behaviorists, however, now confirm the intuitively obvious criticism: that the behaviors that animals commonly perform in
the field when compared from one species to the next are in fact extremely varied
and display themselves (at a least order of complexity) in the form of a spectrum of
behaviors. Yet the field studies have scarcely been taken far enough that the details
of this spectrum are clear, and certainly not far enough yet to quantify such a
spectrum.

In the case of morphology, on the other hand, more detailed considerations have
been known for many years, and a markedly quantitative approach is also available. Even here, although the groups that can be clearly seen in morphological data
when examined at relatively gross observational levels do tend also to emerge when
more quantitative methods are used, so that distinct morphological nodes or peaks
are indeed evident, the greater sensitivity of quantitative morphological methods is
starting to show that such clusters are not entirely discrete and that gradations between one morphological peak and the next exist.

In other words, the level of detail at which we are now able to recognize morphological complexities is such that morphology can be said to display a form that
is quasi-discrete. Behavior, on the other hand (with perhaps exceptions for the most
extreme types), may well display a form that our present investigative techniques
are only able to recognize as quasi-continuous. The nature of the relationship between these two may therefore be considered less an algebraic equation as above and
more as a topological mapping of a relatively complicated type where there is a considerable lack of one-to-one correspondence between the sets. The direction of the
mapping will depend, of course, on what one is trying to suggest from associations
between known behaviors and known structures in living forms. But it is most likely
that we may wish to use such a mapping to deduce something of behavior from
known structures in fossil species of whose behavior we know nothing. It is somewhat unlikely that we would have behavioral information about a particular form
without at least having structural information that is greater by an order of magnitude, so that we scarcely have to consider the reverse mapping. Even in these equations, however, we ought to be cognizant of e (error), in all likelihood rather large,
in the mapping. Such a mapping might well be of the form shown opposite.

Current studies of morphology, especially the ability to quantify morphological
differences, are capable of allowing us to produce a map such as that on the morphological side of the above mappings. The study of behavior, however, is not nearly
far enough advanced to allow us to produce the kind of complexity shown on the
behavioral side of the above mapping. It is exceedingly difficult, therefore, to make
good associations between behavior and morphology in extant species; in those
cases where we have only partially known morphologies (fossils), it is even more
difficult to speculate about possible behaviors. Clearly one way to improve a procedure of this kind is to quantify, if appropriate, the complexities of the behavioral
map itself. A second way of achieving this is to try to perceive the actual interrelationships between the behaviors and the morphologies, and in many instances this
may be partially achieved by looking at the levels of hypothesis between behavior

and morphology. Thus a behavior may be due to several functions, each function producing several forces, and the many forces summating as a resultant stress that acts directly upon the morphology. While the adequate quantification of behavior and the detailed determination of the relevant sublevels for this purpose will be a long time in coming, it may be possible to deduce some of the major steps in the interrelationship and thus lead to a strategy for investigation.

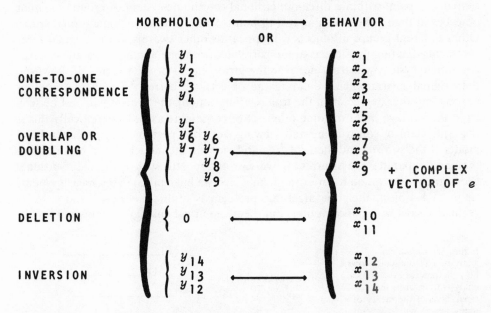

With these ideas about the complexities of the behavior-structural equation in mind, accepting that in reality they are even more complex than supposed here, and accepting that adequate methods for the analysis of behavior have scarcely yet been devised, let us look briefly at the morphometric methods that are available for the detailed analysis of structure in the primates as an Order.

Methods for the Analysis and Display of Structural Variation

Many excellent expository texts are available for supplying understanding, in all its algebraic detail, of the mathematical and computational procedures that comprise the multivariate statistical approach to analyzing morphology as used in these studies (see, for instance, Blackith and Reyment 1971). A number of attempts have also already been made (see, for example, Oxnard, 1973a) to provide nonmathematical pictorial descriptions of the various methods, their relationships to one another, and some of the ways in which they may be used in studies of primate evolution.

However, a brief review of these methods is pertinent here because, for the studies described later in this book, certain additional usages have arisen. In order to understand the questions that are being asked, the results that are displayed and the ideas that can arise from them, it is sufficient to have understanding of these additional uses at geometric levels. If we desire to probe more deeply, and especially if we wish to actually use the methods, we must, of course, apply ourselves to them in a more rigorous manner.

The core of the multivariate techniques is the following. If we suppose that a single object can be defined by three measurements, then that object can be represented as a point within a three-dimensional coordinate system. A group of similar objects will then appear as a swarm of points within that three-dimensional space. Other different groups of objects will appear as other swarms. If the original measurements defining the objects are uncorrelated with one another, then the original coordinate system will truly describe the arrangement of the swarms. If, however, the original measurements defining the objects are correlated with one another to some or other degree, then the true relations among the swarms will best be seen from the vantage point of some other set of coordinate axes. Geometrically this is the equivalent of constructing and viewing from one position a three-dimensional model of the swarms and then rotating and viewing the model from a new position that best separates the swarms. If we can extrapolate such a three-dimensional geometrical description to an example in which we have taken many measurements upon each object, then the algebraic problem is multidimensional and scarcely realizable save with the computer; but the geometrical analogy remains simple (fig.

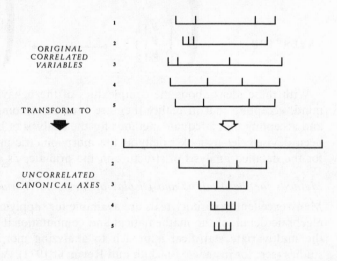

Figure 1. Canonical analysis. This diagram shows how the process of canonical analysis transforms the separations of the means of many groups within several (in this example 5) original correlated variables into separations within several (5) uncorrelated canonical axes. Whereas the arrangement of groups by each original variable is relatively arbitrary with much overlap, the arrangement by the canonical axes is such that the biggest separations are achieved by axis one, the next biggest by axis two, and so on. The statistical significance of separations within any particular canonical axis can be determined. The pythagorean sum of the separations in all canonical axes provides an estimate of separations among the groups that are related to the generalized distances between them.

ORIGINAL
CORRELATED
VARIABLES

TRANSFORM TO

UNCORRELATED
CANONICAL AXES

1). Such methods (and the precise ones that we have concentrated on are canonical variate and generalized distance analyses) are now becoming more commonly used in anthropological studies and form the basis of the approach of this book.

Once multivariate analysis has been performed, however, these techniques may still involve information that is truly multidimensional and that, accordingly, is rather difficult to display. If we are interested in the separation of groups one from another, this can be rather easily examined by study of two-dimensional plots or three-dimensional models of those early canonical axes that contain most of the information. This method of display is relatively simple but loses the information that may be contained within higher axes. Study of many bivariate plots for many higher axes is tedious and does not provide a readily assimilable display.

If we are interested in the nature of the clusterings that are formed by such analyses, then the information can be usefully displayed by constructing the pattern of linkages within the generalized distance matrix. This can be done by applying group-finding procedures (such as single linkage cluster analysis) to these distances. The dendrogram is the visual display that results, but this loses information about the relationships of groups that are not immediately linked to one another.

To avoid these problems, the entire information contained within the total matrix of generalized distances can be examined. But it is most tedious to plough one's way through a complete matrix which may have rows and columns running into double or even triple figures.

Accordingly, we have attempted to display high-dimensional information through other methods. The first of these consists simply of combining the two techniques just described. Thus the minimum spanning tree showing clusterings can be superimposed upon the structure provided by the bivariate plot or three-dimensional model of canonical axes. This allows us not only to see the groupings that are produced but also to get some idea of their relationships to one another. This is an adequate summary as long as it is born in mind that, although the distances of connected groups from one another is correct, the distances of unconnected groups is only a two- or three-dimensional approximation of their propinquity. If, however, marked departures from these approximations are always noted, the method can be of considerable value (figs. 2–6). The second method that we have used is to embed the multidimensional information within an infinite-dimensional space of "functions." For orthogonal independent variates such as canonical variates this can be readily achieved using a sine-cosine function (Andrews 1972). In a totally different way this allows us to "view" groups, intermediates, and outliers within our multidimensional analysis (figs. 7–12).

These methods are of value not only for the examination of a series of known groups but also for investigating the position, relative to known groups, of unknown specimens. Clearly, an unknown can be interpolated into a minimum spanning tree (examples of this are shown in figs. 2, 3, and 4). These various figures demonstrate how problems can arise from the existence of higher dimensional situations. They show a true three-dimensional picture for a series of groups. However,

Figure 2. This figure
displays the difference
between a true three-
dimensional model and its
projection within two
dimensions. Here we have
used the two-dimensional
projection in order to
investigate the relationships
between the groups. Thus the
two-dimensional plot provides
information about directions,
distances, and connectivity.
The directions from one
group to another are of
course correct within the two-
dimensional space. But it can
be clearly seen that the
connectivity is wrong—the
two small groups are not
centrally located, as can be
easily seen by looking at the
reprojection of these
connections into the three-
dimensional model. And, of
course, the distances are also
incorrect. Thus the use of a
two-dimensional plot without
taking account of higher
dimensional information can
provide a totally false view of
relationships.

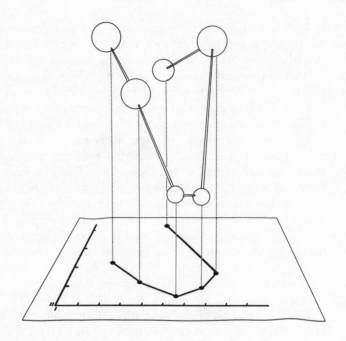

Figure 3. This figure
displays, within the three-
dimensional model, the true
connectivity, and this has
been used to arrive at
connectivity in the two-
dimensional plot. However,
although the plot now
provides both correct
directions within the two-
space and correct
connectivity, the distances are
still incorrect. Thus, although
unconnected, the two small
balls appear, in the two-
dimensional projection, to
be much closer to the others
than in reality they are. This
use of the two-dimensional
plot, although better than that
in figure 2, still may provide
misimpressions about
relationships if not used
carefully.

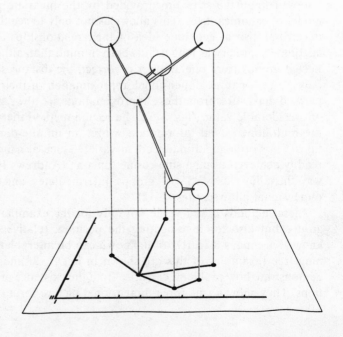

Figure 4. This figure also displays the true connectivity in the three-dimensional model. But in constructing the two-dimensional plot, information about both connectivity and distances has been utilized. The plot below now provides correct directional information within two dimensions together with correct ideas of connectivity and distance. That connectivity is true is shown by the arrangement of the groups being the same for both model above and plot; that distances are true is shown by the two small groups being most distant from the rest in both model and plot. However, we still must live with the limitation that only "connected" distances are correct—others cannot be so. Although in all of these hypothetical examples the analogy is drawn from a three-dimensional model to a two-dimensional plot, the importance of the ideas is greater in the real, complicated situation.

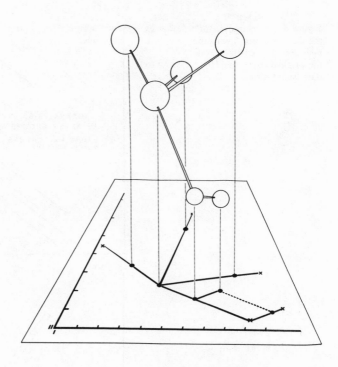

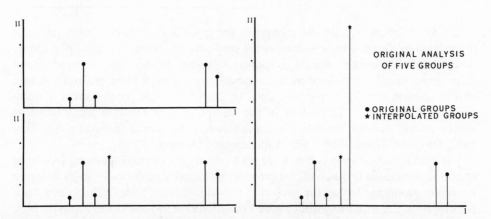

Figure 5. A series of hypothetical plots of canonical axes derived from a study of five given forms (dots) and two unknowns (stars)—one like, the other unlike—a subgrouping of the original forms.

Figure 6. A re-analysis of
the data of figure 5 with the
new axes derived from both
the original (dots) and the
new (stars) forms.

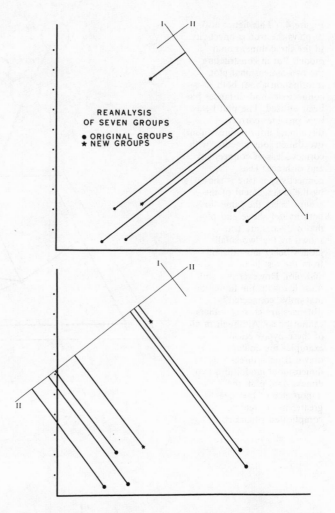

REANALYSIS
OF SEVEN GROUPS
● ORIGINAL GROUPS
★ NEW GROUPS

in the first of these (fig. 2), the groupings are those derived from consideration of
the minimum distance information in the projected bivariate canonical plot alone.
As can be seen from the diagram, as soon as we transpose these links upwards to the
three-dimensional generalized distance model, we realize that an incorrect picture
of the relationships is being provided. In particular, this picture gives a quite
spurious view of the relationships of the interpolated groups (the small balls) as
compared with the real relationships in the three-dimensional model (e.g., see Day
1967, Day and Wood 1968, 1969, and compare Oxnard 1972c).

The next figure (3) shows that, if we know the true minimum spanning tree from
the three-dimensional situation, we can use this knowledge of connectivity to arrive
at correct groupings within the projected two-dimensional plot. In this case we are
passing from the true three-dimensional situation downwards to the bivariate plot.
However we still have an incorrect idea of the relationships in terms of distances.

Thus the two interpolated groups (the small balls) are correctly related to the others but are much too close to them.

If we adopt the methodology of figure 4 however, where both the connectivity and the distances obtained from the true three-dimensional situation are hung upon the skeleton of directions provided by the two-dimensional plot, we obtain a reasonable facsimile of the information. Now we see even in the two-dimensional plot the correct relationships and distances of connected items from one another. In particular we note that the two interpolated groups, represented by the small balls, are both nearest to the central large ball than to any of the others, and yet are still far distant even from that.

Yet another way of understanding some of the problems of interpolation, both indirect and direct, is demonstrated in figures 5 and 6. The first of these diagrams (fig. 5) shows a bivariate plot derived from the analysis of five groups (first frame); a second group that truly belongs with three of the five original groups is interpolated in the second frame; a third group that is truly uniquely different from all groups is interpolated in the third frame. A single axis is able to differentiate the original groups in this simple example and both of the new groups are interpolated into that axis in a position that corresponds with them being close to the three groups at the left-hand end of the axis. However, because we can see both axes at once we know that, whereas the position in the first axis truly represents the position of the first of the interpolated groups, it does not do so for the second.

If the first axis had been used in a biological speculation to suggest, for instance, that items at the left-hand end were bipeds and those at the right quadrupeds, we might possibly be justified in declaring the first of the new groups as a biped. But we would certainly not be justified in suggesting that this analysis shows the second to be a biped. This latter is not like either (e.g., see Day 1967 for an actual example of this problem).

That this is so becomes clearly evident when we study, in a geometrically simplistic way, what would happen if we included the two interpolated groups directly. Figure 6 shows the result. A new first axis would be chosen which would be entirely different from the old first axis. On this new axis, the group which previously was near some of the earlier groups is still close to them. But the group that is clearly unique when we view all the data is also unique in the new first axis. And the new second axis (by definition orthogonal to the first and in this simple geometrical analogy therefore at right angles to it) also maintains the reality of one new group being with the original three groups. That the other now appears linked with the smaller group of two indicates how this information cannot be read independently (even although it is statistically independent) but must be seen in conjunction with the first axis, which declares its uniqueness (see Oxnard 1972c).

Biological speculations about unknown groups can thus only be made when they are truly close to groups in the original analysis. If they are not, then we can say little else than that they are unique. This all seems obvious when we discuss three-dimensional figures and two-dimensional plots both of which we can readily visualize. The procedure of figure 4 is a most valuable display, however, when used to pass from a multidimensional figure to a two-dimensional plot; it then renders a useful simplification of the high-dimensional analysis. The limitations inherent in viewing multidimensional information as in figures 2 and 3 are also clearly demarcated.

In the same way, unknown specimens and groups can be investigated by interpolation into the high-dimensional display obtained from the use of the sine-cosine function (Andrews 1972). Again, as with the previous technique, the analysis is that obtained from multivariate statistics. In this case, the full body of canonical variates (containing all of the information in a multivariate analysis) can be used as the variables $x_1, x_2, \ldots x_k$ in evaluating a sine-cosine function of the type

$$f_x(t) = x_1|\sqrt{2} + x_2\sin t + x_3\cos t + x_4\sin 2t + x_5\cos 2t \ldots.$$

For any given set of canonical variates representing the position of a given group in a k-canonical variate space, the plot of the function is a wavy line (fig. 7) which represents the position of that group. A series of similar groups will be represented by a series of similar wavy lines; a group that is markedly outlying will be represented by a markedly different wavy line; a group that is intermediate will be represented by a wavy line that is intermediate in curvature (fig. 7).

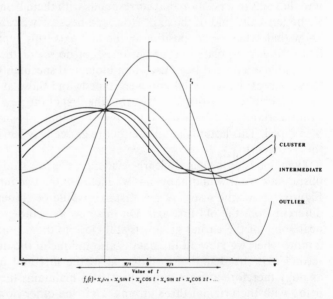

Figure 7. Explanation of multidimensional display by the method of Andrews (1972). Plots, derived as explained in the text, can be obtained for the means of given groups. A cluster of means that are similar in multidimensional space are shown by a set of closely similar wavy lines on the plot. A mean which is widely outlying from the cluster in the multidimensional data-space is shown by a widely different curved line on the plot. An intermediate mean is shown lying in an intermediate position. This method of display preserves variance as a vertical distance on the plots.

CLUSTER

INTERMEDIATE

OUTLIER

Value of t

$f_x(t) = x_1/\sqrt{2} + x_2\sin t + x_3\cos t + x_4\sin 2t + x_5\cos 2t + \ldots$

This method of display can be further understood by examining the way in which each individual component or dimension effects the picture that it provides. This has been done by carrying out a hypothetical study in which all axes except one are held constant and the effect upon the sine-cosine plots of changing that one axis by incremental steps examined. Thus figure 8 shows the effect of altering only the information contained in x_1 of the sine-cosine formula. Because x_1 is not associated with a sine or cosine term, the effect is that the "waviness" of the plot remains constant; the only change is that the entire plot is moved positively or negatively. When

Figure 8. A theoretical study in which the values of x_1 only are changed systematically in order to provide a geometric impression of the way in which the information held in the first canonical variate may be expressed in the sine-cosine plots.

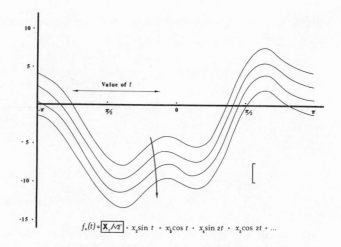

$$f_x(t) = \boxed{\mathbf{x_1}/\sqrt{2}} + x_2 \sin t + x_3 \cos t + x_4 \sin 2t + x_5 \cos 2t + \ldots$$

Figure 9. A theoretical study in which values of x_2 only are changed systematically in order to provide a geometric impression of the way in which the information in a second canonical variate may be expressed in the sine-cosine plot.

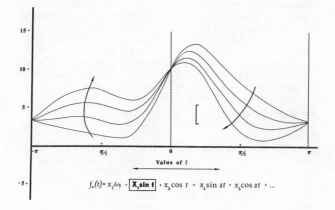

$$f_x(t) = x_1/\sqrt{2} + \boxed{\mathbf{x_2 \sin t}} + x_3 \cos t + x_4 \sin 2t + x_5 \cos 2t + \ldots$$

Figure 10. A theoretical study in which values of x_3 only are changed systematically in order to provide a geometric impression of the way in which the information in a third canonical variate may be expressed in the sine-cosine plots.

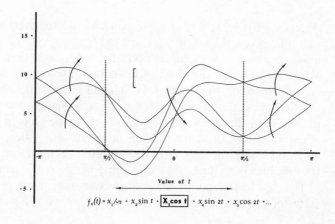

$$f_x(t) = x_1/\sqrt{2} + x_2 \sin t + \boxed{\mathbf{x_3 \cos t}} + x_4 \sin 2t + x_5 \cos 2t + \ldots$$

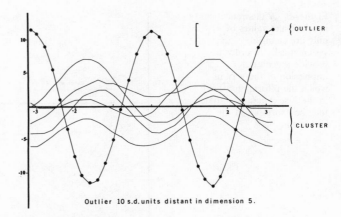

Outlier 10 s.d. units distant in dimension 5.

Figure 11.   In this figure the arrangement of the means of five groups has been displayed using canonical analysis and the high-dimensional display as explained in the text. That the mean of each group is different can be seen from the fact that their continuous curves all lie with considerable distances between them, as judged by the single standard deviation marker given in the diagram. In addition, however, a sixth group (dotted curve) that is actually totally different from the other five is interpolated into the canonical analysis. This new form falls with the others in terms of the early canonical axes which are responsible for most of the separation of the five. But, being interpolated into a set of groups to which it does not belong, the sixth group is markedly different in the fifth canonical axis. The method of high-dimensional display recognizes this and makes it visually obvious through the markedly different degree of waviness of the plot representing the new group. This method of display is thus most useful for determining when an interpolated group (in the actual biological situation, a fossil) does not truly belong with other forms in an analysis.

we examine the effect of changing only x_2, we see that, because a term involving a sine is involved, the waviness is now changed (fig. 9). But the nature of the change is quite regular and is centered upon the values of t that are complete values of π. When we examine the effect of changing only x_3, involving therefore a term that contains a cosine, the waviness is changed in a manner that is centered upon values of t that are at positions represented by $\pi/2$ (fig. 10). Further examples (not figured in this book) show that yet higher axes effect the plots in equivalent ways. It is only when all values of x are varying independently (the normal biological situation when the values of x that are utilized are canonical variates) that we arrive at the complicated wavy lines shown in some of the real examples later in this book.

The problem of noting when an unknown group is really out of place among a series of knowns is especially easily solvable with this method of display. Thus in

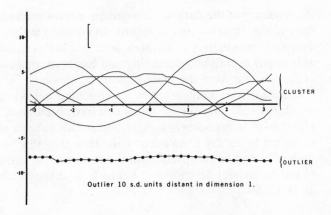

Outlier 10 s.d. units distant in dimension 1.

Figure 12. In this figure the same information as in the previous figure is displayed. Here, however, the interpolated group is included from the beginning in the canonical analysis, and, being more different from the others than any of them are among themselves, the separation of the new group is in canonical axis one. Accordingly, when the sine-cosine curves are plotted, the interpolated group is more obviously totally different from the others. In reality, however, the degree of difference is no greater than in the previous diagram—it is merely more obvious because it involves a canonical axis that produces separation in the first term of the equation $(x_1 \sqrt{2})$ which does not contain a sine or a cosine term and hence separates the plots vertically in the diagram. This compares with the previous diagram in which separation by canonical axis five is dependent upon the fifth term ($x_5 \cos 2t$) and therefore introduces four marked peaks in the curve when plotted from $-\pi$ to $+\pi$.

figure 11 the majority of the groups are separated in two canonical axes, the subsequent higher axes providing no further separation. The addition of an unknown (which is actually different from the other forms because although similar in the first two axes, it differs markedly in the third, fourth, and fifth axes) clearly indicates the nature of its uniqueness. This is the only plot that exhibits an extreme waviness indicative of important high-dimensional difference from the other forms.

When a reanalysis of the multidimensional study is carried out with the unknown specimen included rather than interpolated, then the uniqueness of the unknown may become the earliest information, being separated uniquely in canonical axis one. The separations of the other groups are thus in axes two and three. Again, although in a different way, the sine-cosine high-dimensional display demonstrates the uniqueness of this unknown specimen (fig. 12).

In displays of this type all of the information is presented. However, even here there are limitations, and these devolve upon our abilities to see resemblances and differences between different wavy lines. Although with this method of display the concept of variance is retained (hence the vertical standard deviation marker in figs. 7–12), it is not easy to view more than a small number of groups at one time unless the tendency of the data to form groups is most marked. Again, therefore, we must recognize a limitation in our abilities to visualize multidimensional situations. However, the entire battery of methods described in this section and taken together surely aids in our overall understanding and has been applied in the fairly large number of studies that form the main bulk of this book.

Once again we may return to the structural-behavioral equation. Now, however, we clearly recognize the advances that have been made in understanding the complexities of morphology—it may sometimes take complicated high-dimensional modeling to display them adequately. How may this "sensitivity" of morphological understanding be compared with the crudeness, at current levels of understanding, of our behavioral descriptions? Let us look at the motile behaviors of the primates as an Order.

2   Function and Structure in Primate Locomotion

Primates as Quadrupedal Mammals

Although almost all nonhuman primates are able to perform, to a degree, those nonquadrupedal activities that, in the extreme, characterize a few highly specialized individual species (e.g., almost all primates are able to move bipedally on occasion, and all can swing by their arms), nevertheless these chiefly arboreal animals are primarily quadrupedal in locomotor habit and have in general retained an anatomical format that is correspondingly adapted. Since some type of quadrupedal locomotion must at one time have been characteristic of man's ancestors, it would seem that a thorough understanding of primate quadrupedalism is essential to deriving an understanding of how bipedality may have originated.

In quadrupedal walking and running, the limbs, especially the hindlimbs, provide the necessary vertical and horizontal forces by acting as propulsive levers and struts. When the limb acts as a propulsive lever (i.e., when its longitudinal axis is in front of the limb girdle) the extrinsic retractor muscles (those which run from the body or girdle to the limb bones) are called upon to exert small and relatively slowly applied forces, and in this stage of the locomotor cycle the intrinsic flexor muscles of the limb respond by resisting the tendency of this structure to extend passively. When the longitudinal axis of the limb is behind the girdle joint, the intrinsic extensor muscles running from one limb bone to another are then required to exert relatively small and fairly slowly applied forces in the functioning of the limb as a propulsive strut. If the animal is moving at a constant horizontal speed the horizontal thrust of the limbs need not be much greater, the contact period of the step being used primarily to support the body against gravity. Although the direction of the muscular forces in running are virtually identical to those in walking, running contrasts with walking in the rapidity with which the limbs must be moved relative to the body. Thus, in running, the forces exerted by the muscles in accelerating and decelerating the limbs relative to the body are of greater magnitude and applied with greater speed. To a large extent, the maximum speed of a cursorial animal is determined (a) by the speed at which it can accelerate its limbs forwards when the feet leave the ground after full retraction, and (b) by the speed with which it can accelerate the limbs backward between the moment they reach full protraction and the moment that they are placed upon the substrate. Accompanying such accelerations are the decelerations necessary to reverse the direction of movement making a ballistic pattern of oscillations.

In an arboreal network, running may occur along inclined surfaces, and in such cases the horizontal thrust exerted by the musculature upon the limbs is yet further increased. A wider range of mobility of proximal joints (for placing the limb carefully in relation to an irregular three-dimensional substrate), and distal joints (for placing the cheiridia carefully and safely grasping this complex substrate), is also necessary, but otherwise the muscular demand for arboreal walking and running is qualitatively similar to that in terrestrial animals.

A further modification of quadrupedal walking and running, when carried out in a three-dimensional arboreal environment, is quadrupedal climbing. This term may be used to describe a wide variety of movements. It is possibly best interpreted as referring to locomotion in which the forelimbs reach well forwards or even above the head and pull the body up from above as the hindlimbs, in semi-extended or even extended positions, push the body up from below. The retractors of the proximal elements of both limbs and the flexor muscles (forelimb) and extensor muscles (hindlimb) of the distal elements exert fairly powerful yet not very rapid forces during climbing. In this mode of locomotion, the upward thrust of the body may be achieved by sets of limbs acting together, but the more usual pattern of activity is alternate action. Thus not only are the mechanical requirements on each individual limb quantitatively greater but also there may arise a need for controlling lateral stability during the phase of one-limbed support. Naturally the use of the forelimbs to pull the body up increases considerably the demands on the forelimbs for both power and control, while decreasing, relative to the forelimb, those demands upon the hindlimb.

Morphologically related to these necessarily brief and incomplete functional descriptions, we find that most species of quadrupedal primates carry their bodies on all four limbs of almost equal functional length, so that the trunk is suspended almost horizontally. These animals presumably show relatively little change from what may have been the general ancestral structure of many different mammalian groups. Thus, in contrast to the skeletons of many other recent mammals, all extant primates possess relatively generalized forelimbs with functional clavicles and mobile shoulder girdles, and with the extremity permitted a wide range of movement. Fusion of the radius and the ulna, with consequent loss of rotatory movements of the forearm and hand, so common among many quadrupedal mammals, is not present among the living primates. The hand demonstrates a pentadactyl construction, and although there have been considerable lengthenings and reductions of different individual digits in a number of species, complete loss of any digital ray evidenced in many other mammalian forms is unknown.

The hindlimb, too, is not particularly specialized, even though the pelvis is considerably more fixed than is the shoulder girdle. The nature of the hip mechanism in primates is such as to retain considerable mobility in all three planes of space rather than being structured in relation to a loss of lateral mobility that characterizes the movements of the femur upon the hip in so many highly specialized quadrupedal mammals. Reduction of the fibula and its fusion with the tibia, common in quadrupedal mammals, is found only in *Tarsius* and related fossil species among primates. As with the hand, the foot preserves the five-digit framework without the loss of rays that can easily be shown to have occurred in other orders of mammals.

This basic, rather generalized structure permits almost all nonhuman primates to run, jump, and climb both within the trees and upon the ground.

Variations on the Primate Quadrupedal Theme

At the same time as remaining relatively unspecialized structurally, compared with the mammals as a whole, the primates have evolved within themselves a number of fairly specialized patterns of locomotion; these seem to have been grafted, relatively superficially in most cases, upon the basic quadrupedalism that we have just discussed. This is especially obvious if we view the mechanical demands of these specializations in the light of variations in the structure of the hindlimbs.

Thus in some species an accentuation upon leaping has been added to the generalized running, leaping, and climbing four-footedness described above. This form of locomotion demands rapid, powerful synchronous retraction and extension of the hindlimbs from highly flexed to highly extended positions (see Hall-Craggs 1965). In such species, structural adaptations of the hindlimbs, such as lengthening and strengthening, are closely associated with the locomotor accentuation, although the precise anatomical arrangements vary according to the different types of leaping that exist in the different species. One set of morphological adaptations to a form of leaping is most clearly evident in prosimians such as sifakas and indris and, perhaps, weasel lemurs (figs. 13 and 14); another marked set of adaptations are characteristic of bushbabies and tarsiers (fig. 15 and 16); yet other structural adaptations are present in, for instance, Old World monkeys such as langurs and colobs (Stern and Oxnard 1973; Oxnard 1973c).

Figure 13a. An impression
of *Propithecus verreauxi
coquereli* after Petter.

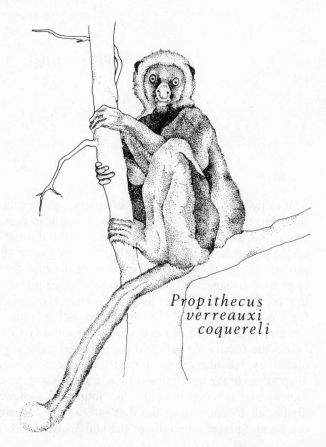

*Propithecus
verreauxi
coquereli*

Figure 13b. An impression
of *Propithecus verreauxi
coquereli* after Attenborough.

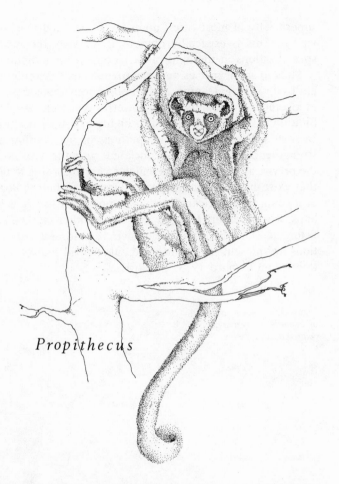

Propithecus

   In other species, rather fewer in number, the locomotor and postural habits include increased suspension of the body weight in part or totally by the hindlimbs.
Again this is an accentuation of some of the normal movements associated with
acrobatic climbing that can be indulged in by almost all primates. Here, in the more
extreme species, associated structures—of the hip—for instance are so modified as
to provide for an even greater degree of three-dimensional mobility than is usual
in more generalized forms. Such adaptation is especially evident in species such as
the howler monkey and the orangutan, but it is also seen, for instance, in some
lorisines (figs. 17 and 18) and the uakari monkey (Stern 1971).
   In some species the arboreal-quadrupedal habit is partially or almost totally re-
placed by reliance on the terrestrial mode of locomotion. Here too, structural
adaptations of the hindlimbs—that is, a slightly greater degree of restriction of three-
dimensional movement in the hip joint and somewhat greater propensities for
digitigrade locomotion in the foot—are found and can be readily associated with
this habit. Species especially of the Old World monkeys, such as the baboons and

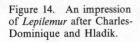

Figure 14. An impression of *Lepilemur* after Charles-Dominique and Hladik.

Lepilemur

hussar monkeys, display limb proportions and structures (e.g., of the hip and foot, Oxnard 1973a; Tuttle 1972) that are not inconsistent with such modifications of the primary arboreal quadrupedal pattern.

Yet another form of terrestrial quadrupedalism is recognizable within the African great apes. Though there are presumably many similarities both biomechanically and structurally with more regular forms of quadrupedalism, it has long been known that the chimpanzee and gorilla have a variety of special structural features that are associated with the knuckle-walking mode. Rather more recently the precise nature of these specializations has been spelled out by Tuttle in a series of publications (e.g., 1969) that define carefully the associated unique myological and osteological features.

In a similar manner, we can look at the structure of the forelimb in relation to these generally quadrupedal locomotor habits of running, leaping, and climbing in the trees. The anatomical condition (short but equal limbs, long body) that is found, for instance, among the tree shrews, and that appears to be characteristic of small four-footed animals in a number of mammalian orders, is not standard among the primates. Grafted on to these habits and structures (apparently) is another series of modifications.

In a number of groups, the forelimb is functioning in modified quadrupedal movements in such a way that it is used for suspending, in whole or in part, the body weight more frequently than in relatively generalized quadrupedal animals. A number of anatomical specializations appear to be associated with this accentuation—

Figure 15.   (a) An
impression of *Galago
demidovii* and (b) of *Galago
alleni* after Charles-
Dominique (the latter based
on a photograph by Devez).

Galago demidovii

Galago alleni

for instance a more mobile shoulder, a lengthened arm, and hook-like fingers. Such
adaptations as these are found in a number of unrelated species, of which the most
obvious are undoubtedly the prehensile-tailed cebids and the arboreal apes.

Again, very many primates occupy themselves in climbing (as compared with
four-footed running up and down almost vertical branches or trunks). Here, although
the forelimb may be used considerably less for actual suspension of the body, it
nevertheless is also subject to considerable tensile force, and associated with this
is an increase in power, both of grasp and of flexion evidenced in its structure.
Among prosimians this appears in two quite diverse groups of species: indriids, in
which acrobatic movements may be almost ricochetal, and lorisines, which are,

Figure 16.   An impression
of *Tarsius spectrum* after
E. P. Walker.

Tarsius spectrum

for some movements, the very reverse of speedy. It is also found among a number
of the more acrobatic monkeys—for example, the proboscis monkey.

Finally of course, while the hindlimbs of almost all primates function almost
exclusively for supporting and propelling the body weight, the forelimbs are free
for much of the time for tactile exploration, for feeding and grasping food and conveying it to the mouth, and for carrying objects (e.g., food, sometimes the young,
and so on). Associated, no doubt, with these last functions are the increased mobility
of the whole forelimb (as compared with the hindlimb) together with, especially,
mobility and control of the terminal elements of the limb: the hand and the fingers.

It is virtually only in man, among living forms, that behavior grossly different
from quadrupedalism is the norm, and it is virtually only man that cannot run
quadrupedally like other primates. His associated unique anatomical adaptations
are well known.

The Less Specialized Quadrupedal Primates

Considerable attention has been paid from time to time to these more obvious
modifications of locomotion and morphological adaptation in the nonhuman
primates as just described. For example, explosive extension and hindlimb architecture as found in the specialized leaping forms such as bushbabies have been well
studied, as has also, for instance, richochetal arm-swinging and the structural correlates in the forelimb such as are found in the brachiating gibbons and siamangs.
It is much less frequently that attention has been paid to generalized quadrupedal
primate locomotion itself.

Figure 17.   An impression of
Loris tardigradus after
Hladik and Hladik.

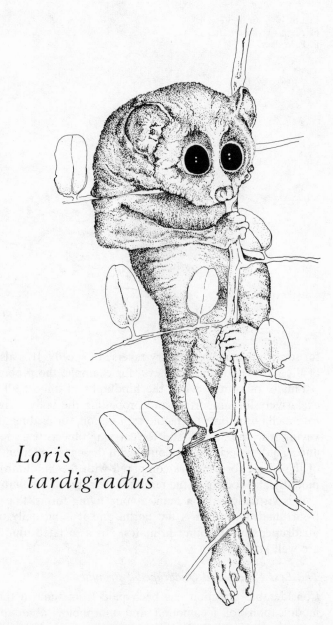

Loris
tardigradus

Once we have identified the rather obviously specialized species, basically quadrupedal though they may be, we are left with a group of species about which rather
little can be said save that they are quadrupedal. These comprise, among the
prosimians, tree shrews,[1] common lemurs, sportive lemurs and dwarf lemurs, and

1. Tree shrews are included here because many authorities group them as primates; there is, however,
   much evidence to exclude them from the order.

Figure 18. An impression of
Arctocebus calabarensis based
on a photograph from
Primate News, February 1971.

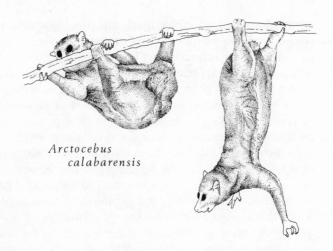

*Arctocebus
calabarensis*

possibly other genera, such as *Daubentonia;* among the New World monkeys, certainly marmosets and tamarins, douroucoulis and titis, capuchins and squirrel monkeys, and possibly other genera; and among the Old World monkeys, certainly most arboreal or semiarboreal cercopithecines and, notwithstanding their considerable leaping powers and abilities for climbing, perhaps even the colobines.

Here, clearly, we are in the borderland of what one worker might judge different enough to mark a separation, yet another would judge as similar enough to disregard minor differences. In other words, the objects of study here are those various primates to which is usually appended the term: "quadrupeds," with usually little else to characterize them save that the extent, speed, and power of running may differ a little, that degrees of climbing ability may vary, as may abilities for nonspecialized leaping, or degrees of terrestrial behavior and so on.

However, there appear to be several major behavioral differences within those primates that are readily observed as primarily quadrupedal; these relate to the differences between the Old World monkeys and the remainder. The movements and postures of the various prosimians and New World monkeys resemble in many ways the quadrupedal species of many other arboreal mammalian groups. But the Old World monkeys differ in relation, not so much to their quadrupedal locomotion as to the fact that when they are at rest, the general bodily posture differs. In Old World monkeys the body is more orthograde so that rather than sitting with the body leaning forwards, or even curled forwards, as is found in many quadrupedal New World monkeys and prosimians, Old World monkeys frequently rest and feed by sitting directly upon the hindlimbs and ischial regions with the trunk in an orthograde position. The forelimbs are thus more completely available for feeding, object manipulation, play, social interaction, and so on.

There is also an obvious locomotor difference that is seen in the comparison of New World monkeys and prosimians, on the one hand, and Old World monkeys, on the other. That is the fact that in terms of locomotion there is relatively little to differentiate the many Old World forms. Though baboons differ from most guenons in being relatively more terrestrial, should a guenon trespass onto the ground, its locomotion does not differ markedly from that of a baboon. Extremes exist, of course; probably no other Old World monkey can imitate convincingly the cheetah-like motion of patas monkeys in full flight. Again, though colobus monkeys leap more often and leap greater distances than guenons and are thus clearly distinguish-able, nevertheless when a guenon does choose to make a large leap, it appears to do it in a manner that resembles, at least as far as we know at the moment, the pattern of colobus monkeys. Among the New World monkeys, in contrast, considerable locomotor differences (as seen on film, in captivity, and in the field) characterize even those species that are most obviously generalized quadrupeds. Although both squirrel monkeys and titis, for example, can be said to be quadrupedal, there is no doubt whatever that there is a marked difference between their behaviors. Even when the one happens to be making the same movement as the other, the manner in which this is done makes easy the differential recognition of the two species. If we look at the most quadrupedal of these animals, the scurrying quadrupedal move-ments that may sometimes be indulged in by marmosets and tamarins, for instance, differ entirely from those shown, for example, by tree shrews when they, too, partici-pate in scurrying, quadrupedal movements.

The morphological correlates of these generalizations have also long been known. Though the various taxonomies of the Old World monkeys indicate a large number of genera, species, and subspecies, they are erected primarily upon the basis of the amazing variety of differences in coat color, in facial shape, and in color and struc-ture of external genitalia, differences that are all visible externally. The general anatomical structure of these species, especially of the postcranium and most particularly of the locomotor apparatus, are remarkably constant (Schultz 1970), notwithstanding certain differences that are present (Oxnard 1967). In contrast, among the New World monkeys and prosimians that are generally quadrupedal, taxonomic differences among genera and easily recognizable clusterings of genera into higher groups are clearly paralleled by anatomical differences in the post-cranium and especially the locomotor system (e.g., Oxnard 1973a).

Locomotor Classifications: Their In-utility

In trying to get a handle on the confusing variety of locomotor activities of primates, many authors have attempted to simplify the picture by making groupings, usually acknowledged to be somewhat artificial, within this complex matrix of behaviors. Although this procedure has been useful for heuristic purposes, and was certainly rather important in earlier stages of investigation when rather less was known about the behaviors of primates, it has become clear, *first*, that the more advanced state of our knowledge denies the reality of grouping, under a given nomen, the overall locomotor activities of *any* set of primate genera. In almost any case that we may discuss, differences between members of putative groups can always prevent their coincident classification. Certainly any attempt to provide such groupings, for what-ever purpose, arouses so much controversy from those who would see in it another purpose, that it is probably valueless. *Secondly*, it has also become clear that there has been much confusion related to the way in which different authors have used classifications. Thus almost all workers who have proposed these terms (and such usage is very old; e.g., see Mollison 1911 and Prieml 1938 prior to Erikson 1963 and Napier and Napier 1967) have been thinking of the locomotor patterns that they believed characterized the movements of whole animals, and they have been making an ordinal overview. Other workers, who do not believe that the locomotion of whole animals can be characterized in this way and who usually have more knowledge of particular species from field studies, have opposed such groupings (e.g., Ripley 1967).

Although Ashton and Oxnard (1964) attempted to use groupings in a different way for the purpose of investigating individual anatomical regions, their classifications have often been misread as applying to the whole animals (e.g., see Ripley 1967; Napier and Napier 1967). The studies of Ashton and Oxnard have explicitly referred, from the earliest publications (1963, 1964), to the fact that their use of the classifications is restricted to the anatomical regions involved, and that, therefore, in studies that progress from one anatomical region to another, animals automatically change category depending upon how particular anatomical regions are used.

The best example of the problems raised by these different usages relates to the unfortunate term "semibrachiator." Many who have used the word "semibrachiator," and who include both the woolly monkey and the proboscis monkey in this group, seem to believe it to mean that these animals carry out an overall form of locomotion that can be called "semibrachiation." Certainly many who have criticized the term have done so upon this ground. This is clearly an unwarranted use of the term, because these two animals do *not* have similar patterns of movement and there is no mode of locomotion that is "semibrachiation." But Ashton and Oxnard have recognized that as long as we confine ourselves to the function of the shoulder or perhaps the forelimb as a whole, there are *some similarities between these two species in comparison with other primates.* Thus, in comparison to patas monkeys, both woolly monkeys and proboscis monkeys use the arm more frequently above the head and in suspensory or tensed postures in locomotion proper, in foraging and in other activities; in comparison to gibbons, both these species parallel one another in doing this much less frequently. Therefore, some facets of shoulder morphology may relate to this functional similarity.

That this applies only to forelimb elements is also clearly recognized by Ashton and Oxnard, who, when looking at function and structure in the hindlimb, recognize that the woolly monkey and the proboscis monkey are almost completely different; the woolly monkey may be grouped with forms such as gibbons, which scarcely ever leap (say, a group of nonleapers); the proboscis monkey may, of course, easily be placed as a member in some group (say, pronounced leapers) that might include mangabeys, although neither of these latter species are as extreme in this regard as some of the very pronounced leaping prosimians (Ashton and Oxnard 1964; Zuckerman, Ashton, Flinn, Oxnard, and Spence 1973).

Because we never have more than a small piece of any higher primate fossil, the functional parallels that can be obtained from regional locomotor classifications may be rather useful in attempting to evaluate the functional correlates of individual fragments. However, since it has not proven possible, in the literature, for this localized use of the groupings to be separated from the clearly unwarranted use of groups of overall locomotion, I have myself in recent years avoided the use of classifications where possible (e.g., Oxnard 1967, 1969a, b, 1972a, 1973a) and have most recently (Stern and Oxnard 1973; Oxnard 1974, 1975) advocated dropping attempts at classifying locomotion. The loss of this tool does not now hamper work in this area because the sensitivity of the analytical methods now available is such that there is no need to group genera into any higher level; whereas univariate and bivariate investigations do not provide enough sensitivity to distinguish features that are relatively similar (e.g., see Robinson 1972), more modern methods involving multivariable examinations are able to make these finer discriminations.

In fact, it is now clearly rather useful to work even at levels below the generic when material permits. Many primate genera contain species that, for one reason or another, may actually move in rather different ways. For example, there seem to be some behavioral and associated structural differences within many of the genera of Old World monkeys (e.g., guenons, mangabeys, langurs—see Oxnard 1967; Oxnard 1973a; Stern and Oxnard 1973; Oxnard, unpublished data) and there are more tentative reasons for believing that this is also true of some New World genera (e.g., howler monkeys—see Stern and Oxnard 1973). In all likelihood, as our knowledge of yet other genera increases, many, perhaps almost all, will be found to provide this more complicated picture.

An Alternative to Locomotor Classifications: Behavioral Aspects

In the absence of providing groupings of genera as a framework against which to place fossil fragments, and recognizing that it may be better nowadays not to provide groupings at all, we must view primate locomotion in some other way. The previous brief descriptions of hindlimb and forelimb activities, and the more detailed information upon which these summaries are based, suggest alternatives that do not commit us to an impossibly rigid or invariant view of any given living primate, yet let us provide an approximate assessment for a particular fossil (Oxnard 1975).

Thus, although we may be rather unwilling or unable to say exactly how the forelimb is used in any given nonhuman primate, we can at least suggest that the major range of differences among all nonhuman primates relates to a broad spectrum that passes from the usage of the forelimb most frequently in the lower quadrant and in compression (perhaps as in the highly terrestrial patas monkey or baboon or in the leaping tarsiers and bushbabies) through to usage of the forelimb more frequently in the upper quadrant and bearing tensile forces as in the highly arboreal gibbon and orangutan. This is not a simple linear sequence; rather, it is a broad swathe or spectrum of possibilities. And we do not attempt to place species at precise points within it, but, rather, provide for a degree of fuzziness for the locus of any individual form. Yet the concept is useful in the study of human evolution when it is recognized that man is different enough from all other forms in not using his arms for locomotion that he lies uniquely away from this broad locomotor band or swathe (fig. 19). Any fossil tending functionally towards man might also be expected to lie outside this swathe.

Figure 19. A broad band or swathe as an analogy for the function of the forelimb in the activities of nonhuman primates. There are no obvious groupings; intermediates may fall at any place within the spectrum outlined by the broad arrow; man is uniquely placed outside the spectrum.

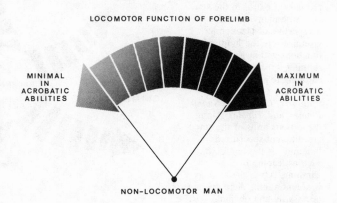

LOCOMOTOR FUNCTION OF FORELIMB

MINIMAL IN ACROBATIC ABILITIES

MAXIMUM IN ACROBATIC ABILITIES

NON-LOCOMOTOR MAN

When we come to view the functional usage of the hindlimb within locomotion, we find that the situation is more complicated. In addition to the function of the hip and leg in generalized quadrupedal locomotion, it is easy to see a number of different extreme modes. There must also be, perhaps less easily recognized, several equivalent series of intermediate forms. Thus the analogy that we might draw for the function of the hindlimb in locomotion of nonhuman primates, is less a simple broad swathe and more a complex starlike shape. Such a descriptive analog allows, for instance, that species with relatively generalized hindlimb functions within quadrupedalism (e.g., squirrel monkeys) may be said to lie within the body of the star and that different species with specialized hindlimb functions lie out along the different arms (e.g., lorisines as clinging forms, prehensile-tailed cebids as tail-swinging types, different kinds of leapers as a series of yet other types, and so on). Such a concept avoids, again, placing animals in groups that may not really exist; it allows intermediate species to be envisaged; it prevents us, as before, from being overly specific in the precise placing of individuals—a degree of fuzziness is allowed (fig. 20). The concept may likewise be useful for human evolution when we recognize that the functions of the human hindlimb are uniquely different from those of any nonhuman primate. Man lies totally outside this star-shaped arrangement of functions for the hindlimb in the same way that, for the forelimb, he lies outside the simpler band-shaped spectrum of functions in nonhuman forms.

An Alternative to Locomotor Classifications: Structural Correlates

We may now ask the question: to what extent do the anatomies of the forelimbs of extant, nonhuman genera reveal evidence of a broad structural spectrum or swathe (ranging perhaps from the architecture of forelimbs in terrestrial quadrupedal monkeys to those in brachiating gibbons) and parallel the functional spectrum just described? Previous multivariate morphometric studies of the pectoral girdle in

Figure 20.   A star-shaped
spectrum as an analogy for
the function of the hindlimb
in the activities of nonhuman
primates. Central in the star
are those many species,
irrespective of taxonomic
group, where the hindlimb
functions within generalized
running, springing, and
climbing (e.g., *Aotus,
Callicebus, Callimico,
Callithrix, Chiropotes, Saimiri,
Phaner,* and *Tupaia*). Along
the various arms of the star
are different specialized
functions of the hindlimb,
such as leaping or slow-
climbing. Again the analogy
is a continuum; there are not
necessarily any obvious
groupings; intermediate and
incipient forms may exist.
Man is outside the analogy.

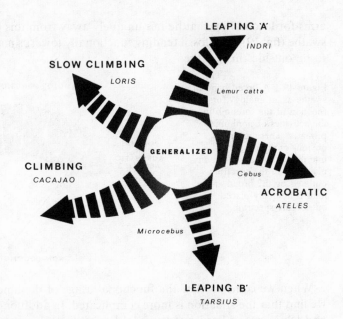

primates (based on seventeen measurements on each of 376 specimens of 28 genera
of primates as summarized in Oxnard 1973a) show clearly that shoulder structure
among nonhuman primates is rather readily correlated with this idea (fig. 21). These
studies also show that the shoulder of man is organized structurally quite differently from that of the nonhuman primates; structurally as well as functionally the
human shoulder lies uniquely outside the spectrum of nonhuman species (fig. 22).

It is of considerable interest that this structural-functional association is not confined to the shoulder region (Oxnard 1975). Using multivariate statistical methods,
study of a wide range of dimensions of the forelimb as a whole, presents a generally
similar picture. In this case a series of measures of the relative dimensions of the
forelimb (relative shoulder breadth, relative upper limb length, brachial index, relative hand length, relative hand breadth, and relative thumb length) are available
taken from a total of 366 specimens representing twenty-six primate genera from the
work of Professor A. H. Schultz. The arrangement of primate genera obtained by
multivariate study of these data is presented in figure 23, which demonstrates a
broad swathe of nonhuman forms that extend, at one end, from extreme hopping
and terrestrial forms in which the forelimb is much less involved in locomotion or
shares only equally with the hindlimb, to, at the other limit, specialized brachiating
species in which the forelimb is dominant over the hindlimb, being subject to almost
the entire forces of locomotion. As with the studies on the shoulder girdle, man is
uniquely separated from the nonhuman spectrum.

That these two arrangements (of the shoulder girdle alone and of the entire forelimb) are likely to be representing similar structural-functional associations is suggested by the remarkable concordance of the rank order of genera between the two
studies (fig. 24). Thus, of the many primate genera available, only two (patas

Figure 21. These bivariate plots of canonical axes one and two and one and three respectively are those obtained from the analysis of seventeen dimensions of the shoulder girdle (Oxnard 1973a). They display that in these axes the primates as a whole occupy a three-dimensional band-shaped arrangement. Squares = hominoids (the circle provides a single standard deviation unit around the position of man); diamonds = New World monkeys; dots = Old World monkeys; triangles = prosimians.

Figure 22. This bivariate plot of canonical axes one and four from the same analysis that gave rise to the previous two plots demonstrates that man is uniquely separated from the nonhuman primates. These latter are scarcely separated at all by this and all later axes. Conventions as in figure 21.

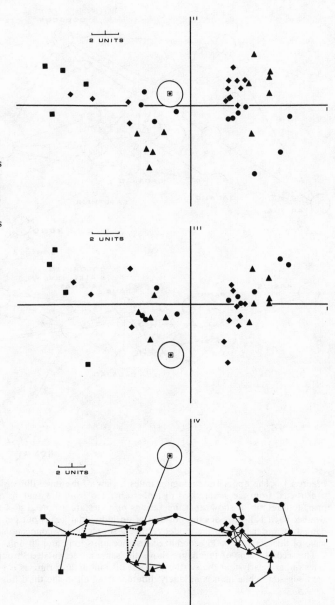

monkeys and baboons), are markedly differently placed when we examine the first canonical axis of each of the two studies. And even here, examination of higher canonical axes for overall forelimb proportions does suggest that these two exceptional genera are indeed outlying forms, as they are in the study of the shoulder girdle.

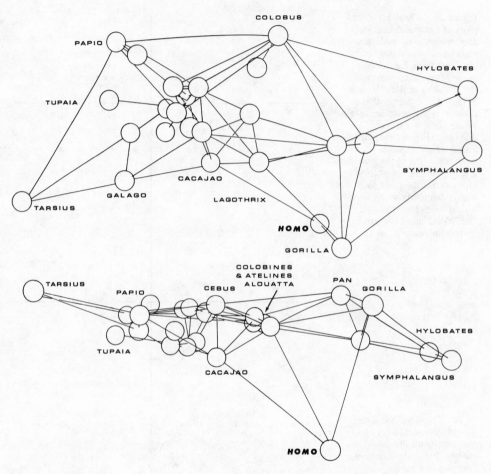

Figure 23.   The upper frame demonstrates a view of the three-dimensional model that can be constructed from the generalized distances obtained from the analysis of eight dimensions of the primate forelimb. It shows that the various primates are arranged as a two-dimensional band or swathe. Each ball provides the position of an individual genus but peripheral genera only are named. In order to keep the model relatively simple, many connections are omitted. The general scale of this diagram is 36 generalized distance units long and 20 units wide.

The second frame of this figure shows the same model rotated through ninety degrees. It indicates rather clearly that the swathe or band of nonhuman primates is generally flat and is here seen edge-on, while man is uniquely different from all in the third dimension.

The one region of the forelimb that has not been examined closely in this way is the structure of the hand. However, it seems clear that the functions of the hand are so much more complicated than those of the forelimb as a whole or of the shoulder alone, that it is unlikely that the hand would follow this simple set of associations. Pending the prosecution in this way of both functional and structural studies of the hand, we may well guess that it would present a far more complicated picture. But at least in those studies so far available, it would seem that the notion

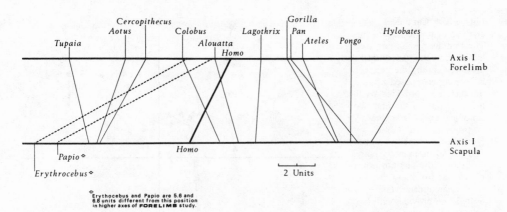

Figure 24.   In this figure the positions of certain named genera are indicated for the first canonical
axes of both the study of eight forelimb dimensions and nine dimensions of the scapula. The
arrangement of the genera are similar in the two studies, so that most of the joining lines in the
diagram are generally vertical. Such translocations as do occur are confined to reversals within
adjacent localities. One major difference is that baboons and patas monkeys (*Papio* and *Erythrocebus*) are outliers in the scapular study but are central in this diagram for the forelimb study. This
latter impression is spurious and is corrected by examination of higher axes in the forelimb study.
In general both investigations provide the same information about the arrangement of the primates.
The marker indicates the scale of the diagram in standard deviation units.

of a simple spectrum of function in the forelimb as a whole is clearly reflected in a
simple spectrum of structure in that same member.

We may also ask, as with the forelimb, if there is any structural information about
the hindlimb that provides reality to the analogy of the star-shaped arrangement of
functions postulated in the previous section for this anatomical region (Oxnard
1975). Interestingly enough, although the hindlimb has perhaps been less fully
studied by these methods than has the shoulder, data on the pelvis in lower primates
(fig. 3 of Oxnard 1973c) does suggest that such an association may exist. This study,
again using the multivariate morphometric approach, of a wide range of dimensions of the primate pelvis presents a generally similar picture. In this case a series
of measures of the pelvis (four dimensions related to relative positions of joints,
five associated with muscular positions and leverages taken on 430 specimens representing 41 genera of primates) are available from the studies of Zuckerman, Ashton,
Flinn, Oxnard, and Spence (1973). The arrangement of numbers of primate genera
obtained by this study is generally star-shaped when synopsed as a minimum spanning tree in two dimensions (Oxnard 1973c; fig. 25). The species central to the star
are generalized quadrupedal forms; those outlying in the rays of the star are species
that are extreme in the function of the hip in locomotion in the variety of different
ways outlined previously. Man lies uniquely outside the star-shaped arrangement.

That this particular functional-structural correlation is not peculiar to the pelvic
region is suggested by further study of a series of relative dimensions of the hindlimb (relative hip breadth, relative lower limb length, crural index, relative foot
length, relative foot breadth, and foot length relative to lower limb length taken on

Figure 25. Canonical
analysis of pelvic dimensions.
The minimum spanning tree
based upon the generalized
distances among the genera
is here superimposed upon
the directions provided by the
bivariate plot of the first two
canonical axes. Triangles
= prosimians; diamonds =
New World monkeys; dots =
Old World monkeys; squares
= hominoids. The position of
man is far from any
nonhuman species. The
dotted line provides only
one of many near-minimum
links between man and the
remaining primates. The
generally star-shaped
arrangement of the genera is
clear, with man falling well
outside the star. The precise
positions of named genera
in this study are better seen
in figures 38 and 39,
chapter 3.

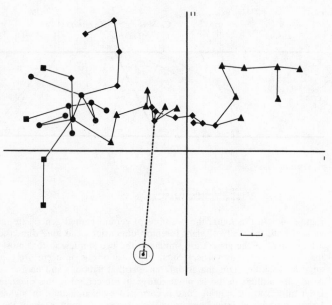

a total of 440 specimens representing twenty-seven primate genera), again available from the studies of Professor A. H. Schultz. Multivariate analyses of these
dimensions produces a set of structural relationships that approximate to a more
complicated star-shaped arrangement of nonhuman genera, with, again, man lying
uniquely well away from all (fig. 26). Thus, of the nonhuman primates, many relatively generalized quadrupedal monkeys (of whatever taxonomic group) lie together
as a central body. Each of the arms emanating from the central body contains genera
such that functionally similar extreme and intermediate species lie together (again
irrespective of taxonomic groups). One such arm proceeds, for instance, from the
central group through dwarf lemurs through bushbabies to tarsiers—one morphological spectrum of "vertical clingers and leapers." Another, quite different, arm
proceeds from the central group via ring-tailed lemurs to the various indriids—a
second morphological spectrum of "vertical clinging and leaping" species. The slow,
clinging lorisines form yet another such arm, as do, separately, the prehensile-tailed
cebids, the curiously acrobatic pitheciines, and the unique genus *Daubentonia*. All
of these species have quite different functions of the hindlimb in locomotion; all
exhibit structural differences that are associated with this; a sufficient number of
the arms of the star contain genera from more than a single taxonomic group so
that we are not picking up merely coincident taxonomic similarities (though no
doubt taxonomy is inextricably involved in this picture).

And again, that these two structural arrangements (pelvic and total hindlimb
proportions) are likely to be representing similar structural-functional associations
is suggested by the remarkable concordance of the rank orders of genera along the

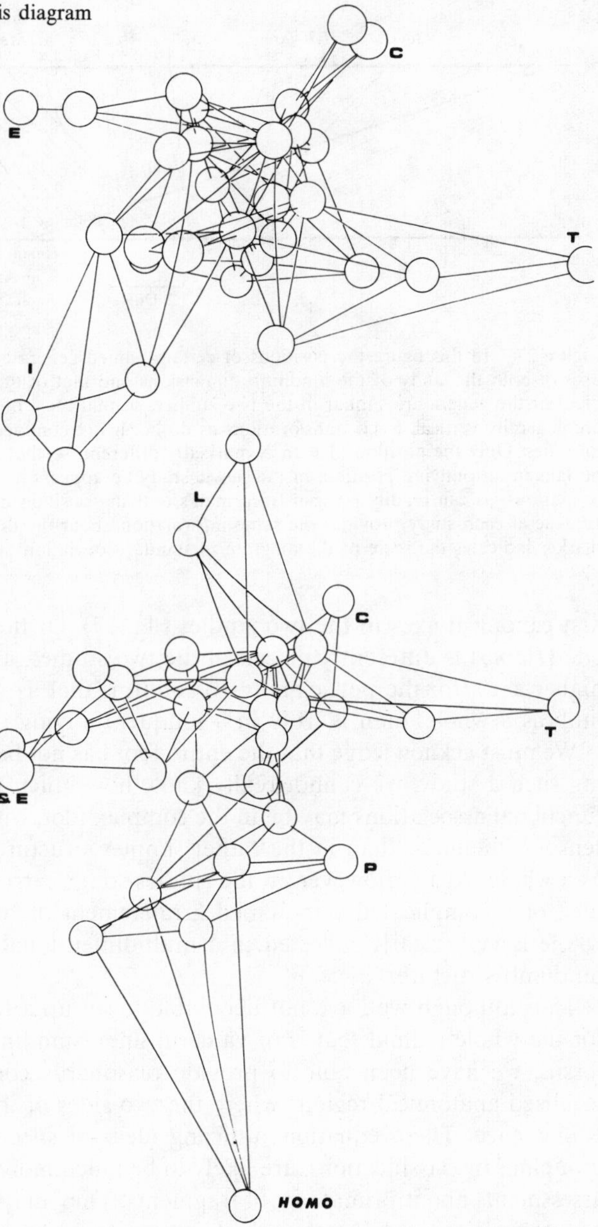

Figure 26. The upper frame of this diagram provides a view of the three-dimensional model that can be reconstructed from the generalized distances obtained from the analysis of six dimensions of the primate hindlimb. They show that the various nonhuman primates are arranged in a rather complicated star-shaped formation. Each ball provides the position of each individual genus. Those genera placed centrally in the star happen to be those that can be designated as basically quadrupedal in locomotion (e.g., capuchins, *Cebus;* squirrel monkeys, *Saimiri;* marmosets and tamarins, *Callithrix* and *Leontocebus;* macaques, *Macaca;* and cercopitheques, *Cercopithecus*). That ray, or arm, of the star marked "T" comprises, at the extreme, the leaping tarsiers (*Tarsius*) and, more centrally, the bushbabies (*Galago*), connecting with the quadru-pedal nucleus through mouse lemurs (*Microcebus*) and dwarf lemurs (*Cheirogaleus*). The arm marked "I" contains the leaping Madagascan indriids (*Indri, Propithecus,* and *Avahi*), and they link with the central quadrupedal nucleus through various lemurs (*Lemur* and *Hapa-lemur*). The outlying arm marked "E" includes the terrestrial quadrupedal baboons and patas monkeys (*Papio* and *Erythrocebus*) linked to the centrally lying macaques, mangabeys, and cercopitheques (*Macaca, Cercocebus,* and *Cercopith-ecus*). The arm marked "C" includes the curiously acrobatic uakari and saki monkeys (*Cacajao* and *Pithecia*), and they link with the centrally lying titi and squirrel monkeys (*Callicebus* and *Saimiri*).

The lower frame demonstrates the arm "P," comprising the various acrobatically inclined apes, hidden above; man, though linking most closely with various apes, lies in fact outside the star-shape provided by the other genera, including the apes. The general scale of this model is 30 distance units long and 24 wide.

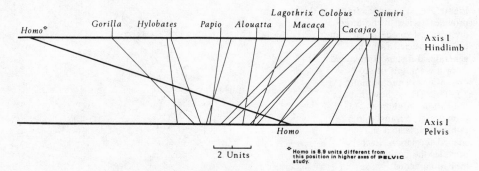

Figure 27.   In this figure the positions of certain named genera are indicated for the first canonical axes of both the study of the hindlimb dimensions and that of the pelvic dimensions. The arrangement of the genera are similar in the two studies, so that most of the connecting lines in the diagram are generally vertical. Such translocations as do occur are confined to minor reversals within adjacent localities. Only the position of man is markedly different, so that although in the hindlimb study he falls in an outlying position, in the pelvic study he appears to lie rather centrally. This impression is spurious, as can readily be seen from man's outlying position in higher axes of the pelvic study. In general each study provides the same information about the distribution of the primates. The marker indicates the scale of the diagram in standard deviation units.

first canonical axes in the two studies (fig. 27). Of the many primate genera, only one (*Homo*) is differently placed in the two studies, and even here examination of higher axes (for the pelvic analysis) confirms that as with the analysis of the hindlimb as a whole, man is truly in a unique, outlying position (see also fig. 26).

We must acknowledge that the entire foot has not been studied in this way. Pending such a study, we cannot really know how much more complicated structural-functional associations may be in the complex foot, with its additional grasping and sensory functions, than in the rather simpler structure of the pelvis or of the limb as a whole. Again, however, in the studies so far carried out it would seem that the idea of a complicated star-shaped arrangement of function in the hindlimb as a whole is very clearly reflected in a multidimensional star-shaped arrangement of hindlimb structure.

Thus although we have not been able to set up a functional-structural equation for the whole animal that is of value in understanding the shapes of fragments of fossils, we have been able to provide reasonably convincing equations for more localized anatomical regions where the two sides of the equation show their rough equivalence. These equations utilizing ideas of spectra and continua rather than groupings or classifications, are likely to be much more useful in helping us to make assessments about primate fossil fragments. They may well allow easier recognition of intermediate and even incipient functions in both extant and fossil species, and with these we may more readily associate equivalent intermediate or incipient structural adaptations. They may also allow us to detect uniqueness in a fossil when it is present. Let us now examine a series of anatomical regions for which fossil fragments are available and in which the multivariate approach has been used to display the structural side of the equation.

The Shoulder and Pelvis

Diversity and Uniqueness in Human Evolution

One of the biological problems that specifically interests us is that of providing, if possible, some estimate of the functional associations of those particular fossil remains that have been classified from time to time as australopithecine. This particular problem is of especial interest and difficulty because these fossil remnants are thought to be closely related to human evolution. They display features (especially of the teeth and skull) that seem to link them closely with man. Their post-cranial elements must, therefore, be viewed against the background of the unique series of specializations characteristic of man and related, in locomotor terms at least, to functional-structural adaptations for upright posture and gait.

On the other hand, the fossil fragments should also be viewed against the background of the diversity of locomotor patterns that are known among extant non-human primates. Possibly at almost all times during the evolution of the primates there have been species variously adapted for quadrupedal running, leaping, and climbing in the trees, and in many of these it is likely that specializations with emphases on the hindlimb (e.g., such as is involved in extensive leaping: *Galago, Colobus*) and on the forelimb (e.g., in acrobatic climbing activities: *Lagothrix, Pongo*) have broadened the diversity of quadrupedal forms into the complex arrangements described in the previous chapter. It is from one or another part of this diverse quadrupedal mosaic that human bipedalism has been derived.

Such a series of comparisons is rarely attempted, since it is commonplace for investigators to compare the australopithecines only with living man and African apes.[1] Even the inclusion of pongids in such comparisons is usually either to confirm resemblance to man and deny similarities to nonhuman species, or to seek similarities with specific nonhuman forms (e.g., the African great apes) for the purposes of bolstering (rather than testing) previously determined genetic hypotheses. It is somewhat less often that investigators make comparisons with a wide spectrum of primate species in order to determine relative degrees of similarity, parallelism, and divergence, and especially in order to attempt to deny hypotheses, the most valid point of scientific studies.

It is possible to question pooling the australopithecines into a single group in these studies. However, when we recognize the extraordinary variability that exists within living primates at the generic level, we realize that this is a reasonable exercise. For the differences that are known to exist among the individual fossils (for instance, between the robust and gracile forms) are not nearly as large as those found within some living genera. Comparisons may be made, for instance, with the genus *Homo,* who displays continuous variations in height within living man that greatly exceed any that are known to have existed among the australopithecines. In comparisons with the genus *Pan,* the discrete differences between the postcranial elements of the common and pygmy chimpanzees are almost as great as—indeed, may be greater than—those among specimens of adult australopithecines. In comparisons involving yet other genera, even bigger differences may exist. We may

1. The studies of Straus (e.g., 1949) are a notable and elegant exception.

compare, for instance, the marked environmental differences found within the genus *Cercopithecus,* among which talapoin monkeys inhabit low bush and diana monkeys live almost exclusively in the high canopy. Even in some genera large locomotor differences may be found; some langurs are largely terrestrial, others mainly arboreal; some bushbabies are extreme hopping forms, others are primarily quadrupedal runners. For the discussion of results it is entirely reasonable to consider the fossils together in comparison with such generic groupings of extant species.

A second rather practical reason why it may be useful to consider these fossil fragments in this way relates to the fact that morphologically they tend to fall rather close to one another. It is true that there are considerable differences among the various skull fragments, but certainly not much more than can be seen among the various macaques, for instance, and certainly less than that existing between the sexes of chimpanzees and of gorillas, for example. And in those few instances where equivalent postcranial regions can be compared—and the talus is perhaps the best example here (see chapter 4)—the differences between the different fossils are actually rather small. This is especially so when they are viewed in the light of differences within and between various extant genera.

This discussion does not deny, of course, differences within the group, and in the statistical manipulations the individual specimens are not pooled. But for the purposes of discussion of the entire subfamily, such grouping would seem useful at the present time. For although there are some workers who would place those various fossils at the generic level under separate names—*Paranthropus, Australopithecus, Zinjanthropus,* and even *Homo* as in *"H. habilis"* and *"H. africanus"*—many others would rather more conservatively place the entire set in the genus *Australopithecus* and then attempt to see if there is rationale for further splitting within the genus.

The questions, then, that we are interested in asking about any postcranial fragment of the australopithecines stem from the discussion of the previous chapter and are the following: To what extent does the morphology of the australopithecine fragments indicate (a) close functional similarity with extant *Homo sapiens* (i.e., similarly uniquely bipedal), or (b) close functional similarities with any of a wide diversity of essentially quadrupedal forms as represented today by the living nonhuman primates, or (c) a pattern of functional associations that point to intermediacy between man and one or another part of the nonhuman matrix, or even (d) a pattern of functional associations that may be unique to themselves (a possibility rather less frequently considered)? Answers to any of these questions would provide useful and interesting data for primate evolution, although we must bear in mind that, for australopithecines, the first possibility is generally believed to be the case (e.g., Le Gros Clark 1964; Robinson 1972).

Initial Studies of the Shoulder

Our own laboratories have provided some initial studies of the shoulder region in primates that are especially pertinent to these locomotor questions. The method utilized in those studies attempts to draw associations between the biomechanical (locomotor) functions of the shoulder in a wide range of primates and the detailed muscular and osteological structure of the shoulder that they display. For example, the arrangement of the upper fibers of the trapezius muscle as they attach to the spine of the scapula is such that they are directed cranio-caudally in those species in which the girdle tends to be moved more in the cranio-caudal axis during the relatively pendulum-like movements of the forelimb in the propulsive and retractive strokes of normal quadrupedal locomotion. But in those species in which the arm is frequently used above the head during acrobatic climbing or even in brachiation, the cranial fibers of trapezius tend to be directed medio-laterally towards the spine of the scapula, in which position they are biomechanically better able to provide one of the forces that go towards rotating the scapula, a movement vital to arm-raising and producing a wider excursion than is usual in regular quadrupedal movement.

Functional-muscular associations such as these, then, give rise to metrical definitions of aspects of the form and proportions of the scapula, clavicle, and humerus. For instance, in relation to the example quoted, whether or not the fiber direction of the upper part of the trapezius is cranio-caudal or medio-lateral seems to be related to whether or not the scapular spine is directed medio-laterally or cranio-caudally respectively. In each case, it is most efficient to have the insertion of this part of the muscle into the bone at an angle as close as possible to the normal. Although this association cannot be complete—that is, other biomechanical and, indeed, possibly other features may interfere with the relationship—it at least portrays to a considerable extent the actuality of the situation. Such an osteological feature can be rather readily defined in a metrical manner and thus quantified.

A series of features are available that have been chosen in this way and the data thus produced are examined by the multivariate statistical method. This method attempts to add together the information inherent in such a suite of measurements while at the same time making due allowance for the redundant information that must be contained within them (in statistical language, for the degree to which the various dimensions are correlated). Shorthand descriptions of these methods have been provided in the first chapter. There are many problems with studies of this type. One concerns the basic form of the data: should we utilize raw measurements, or should ratios, the traditional tool of the physical anthropologist, be used? Should we attempt to make alterations of the data based upon ideas of including, or excluding, those parts of the information that can be related to differing absolute sizes of animals? Should mathematical transformations be applied to take account of the statistical non-normality that is bound to be present in incomplete and non-randomly sampled data, and so on. Thus, for example, the requirement, imposed by the necessarily small samples that are available for most postcranial studies of primates, that sex subgroups, age ranges, and subspecific and specific divisions be pooled raises distinct problems as to the appropriateness of using multivariate analytical methods. Accordingly, in all of our own studies we have attempted to carry out series of preliminary univariate and bivariate tests, and in many cases even preliminary multivariate analyses, in order to understand the degree to which the data conform to the theoretical forms required for absolute rigor in multivariate analyses. Such features can be disturbing, especially in studies where specimens and groups are so close together that significance testing is an important part of the procedure. This is especially the case in those studies investigating racial, subspecific, or specific differences. However, in our own studies the groups are usually so different from one another that the fact of difference is not at issue. Significance testing is thus no problem, and under these circumstances the multivariate statistical method provides much greater robusticity to possible aberrations of the data set such as arise from the above mechanisms. A detailed study of these and other problems is presented in Oxnard (1973a).

Suffice it to say that the results of these studies of the shoulder girdle have been most useful in demonstrating the marked degree of inductive detail that may some-times be provided about functional-structural relationships by the multivariate morphometric approach. They have shown, for instance, that while conventional investigations have supplied considerable information about the relationship be-tween the increasing mobility of the shoulder and differences in its musculo-skeletal framework as seen in the sequential examination of prosimians, then monkeys, and finally apes and man (Frey 1923; Miller 1932; Inman, Saunders, and Abbott 1944), the multivariate method of analysis and display allows (a) a better overall view of the many different facets of shoulder shape that exist, (b) a more detailed under-standing of the spectrum of functional relationships among individual primate genera and (c), most usefully, the uncovering of aspects of structural-functional relationships in the shoulder that are quite unsuspected from the more classical studies (summarized in Oxnard 1973a and further discussed in Stern and Oxnard 1973).

Thus certain aspects of the shape of the shoulder girdle—for example, the cranio-lateral twistedness of the scapula—seem to be readily associated with the degree to which the shoulder as a whole bears compressive forces as in regular quadrupedal activities or, to an increasing extent, tensile forces as in a variety of more acrobatic activities culminating in the extreme acrobatic behaviors of the orangutan and the brachiation of the gibbon and siamang. Other aspects of the shoulder girdle—for instance, in terms of scapular shape a medio-lateral (dorso-ventral) compression—seem to be related to the general mobility of the shoulder girdle in three dimensions, the girdle being medio-laterally (dorso-ventrally) increased in those more strictly cursorial species (ground or large-branch running animals), and reduced in those arboreal species in which climbing within a three-dimensional small-branch environ-ment becomes more and more a preferred activity. And this second relationship seems to apply irrespective of the associations of nonhuman primate genera in terms of the first morphological spectrum. In other words, the various nonhuman primates are arranged morphologically in a broad swathe. But in that one species for which neither of these modes could possibly apply (*Homo*, who neither uses his arm for locomotion in a tensile-compressive situation nor is involved in any terrestrial-ar-boreal habitat) the finding is that, although man has a projected position within this part of the result, additional information indicates that his true position is uniquely different. This presumably correlates well with the unique (nonlocomotor) functions of the human forelimb and shoulder (see also chapter 2 and figures 21 and 22).

The Shoulder of Australopithecine Fossils

In addition to this elucidation in extant species, these studies of the primate shoulder are also applied, although in only the most tentative way, to certain fossil fragments of australopithecines (a portion of scapula from Sterkfontein, most of a clavicle from Olduvai). This study has suggested that the conclusion that the australopithecines are essentially similar to man is not, at least for the shoulder girdle, valid. Thus when viewed in three different but related studies—(a) based simply upon the morphometrics of the fossil fragments, (b) using the idea of a nearest analog, and (c) based upon the concept of minimum morphological pathway (Oxnard 1968a, 1968b, 1969b, 1973a)—the shoulder of the australopithecines is seen to be uniquely different from that of man. The results indeed go further; they suggest that although the fossil shoulder is also different from that of most nonhuman primates, it does possess an analog in the form of the shoulder of the orangutan—the shoulder of an acrobatic tree-climbing species.

> These findings do not, of course, mean that there is any actual close genetic affinity between man and the orangutan. What they very well could mean is that the presumed common ancestor of man and the African great apes is an arboreal creature capable of considerable acrobatic activities in the trees, and possessing, for that reason, a shoulder morphology more like that of the present-day orangutan than that of the gorilla or chimpanzee. (Oxnard 1973a)

But these findings are so tentative and surprising, are based upon such fragmentary fossils (the scapula is already more incomplete now [Robinson 1972] than it was when first figured by Broom, Robinson, and Schepers 1950), and depend upon such incomplete and reconstructed data, that they arouse in themselves little more than a flicker of interest. Clearly a more careful study of a number of anatomical regions is required before any sound evidence, positive or (equally important) negative, is forthcoming.

However, since those multivariate studies on the shoulder were performed, a number of other anatomical regions have been investigated using somewhat similar methodologies. And though the number of these is not large and does not by any means scan the entire primate postcranial skeleton, yet there are now sufficient such studies to make an overview at this time a useful procedure.

These newer studies, both of our own and of other workers, are beginning to point in directions that, while not necessarily denying close genetic relationships between these fossils, the African great apes, and modern man (comparative morphometrical osteology does not provide unequivocal information about such matters), nor yet conveying information about the temporal propinquity between these species (again the comparative morphological method does not help us much here), are nevertheless suggesting that, in terms of functional adaptations of postcranial skeletal structure, similarities of the australopithecines to modern man and his nearest extant relatives, the African great apes, are not always obvious. Other living genera may provide better functional analogs. Thus it is that these African fossils may indeed be pointing to new questions for human evolution.

The Evidence of the Pelvis

That initial study on the shoulder is now followed by a somewhat similar type of analysis of pelvic morphology (Zuckerman, Ashton, Flinn, Oxnard, and Spence 1973) in which the same approach is applied, although including a number of methodological refinements not available at the earlier times. Again, the study first attempts to make, from the behavioral and functional information that is available, some simple characterization of the differences in biomechanics that result from the functions of the pelvic region in locomotion throughout the primates. The investigation secondly tries to reveal the soft tissue anatomy (principally of muscles and joints) of the pelvic region and relate them to the biomechanical differences. Thirdly, the investigation points to osteological features associated with the pattern of muscles and joints, attempts to define them metrically, and combines the resulting information by means of multivariate statistics such as canonical variate and generalized distance analyses. Fourthly, the information contained therein is demonstrated, using such methods as canonical variate plots, clustering algorithms, minimum spanning trees, and high-dimensional displays. Information is thus provided to us about the way in which the different genera of the primates are separated by the morphological data, and we can assess the degree to which such separations are associated with the initial functional suggestions about primate locomotion obtained from the surveys of literature, viewing of films and observations made by colleagues. All of this forms a background into which data from an australopithecine innominate bone may be interpolated, and through which some answer about the evolution of function in the fossil may be arrived at.

In the case of the investigations on the shoulder girdle, although it was appreciated that the information ought to be most valuable in the assessment of fossils, the original work was not undertaken, in the period 1959 through 1964, with any specific fossil remnants in mind. This was because the methodology was sufficiently new that it required to be tested solely within the ambience of living forms; it was thus hoped that the approach might be judged in its own right as a method of investigating skeletal form, without acceptance of the methodology being clouded by the dust that is always stirred by inferences about human and prehuman fossils.

It was for this reason, therefore, that, several years later in 1968 and 1969, when fossil information was brought into the shoulder studies, the interpolations were awkward and bedeviled by the problems of missing and incomplete data. Thus the speculations about the shoulder fragments at Sterkfontein and Olduvai are somewhat less than satisfactory and have to be taken with a grain of salt, it never having been the original intention to examine these fossil fragments in this way.

In the case of the more recent study on the pelvic girdle however, the investigations are aimed from the beginning at the analysis of the best of the available australopithecine finds (from Sterkfontein). A cast of the specimen for the right hand side was made available to us some years ago through the courtesy of Professor John T. Robinson. Although such a cast can never provide the full details of surface markings, its dimensions were checked against the original by Professor Robinson and deviations were shown to be less than 0.5mm. The Sterkfontein innominate is not complete, and certain small reconstructions have been necessary. One of these (the region around the anterior superior iliac spine) utilizes information from the opposite side, where the region is well preserved, and this reconstruction, also provided by Professor Robinson, seems very acceptable. The second interference with the fossil has been to straighten out the acute bend that exists in the pubic part of the bone. Here the matter is somewhat less easy, but it has been attempted by utilizing the information that there is, in all primates, a smooth and uninterrupted curvature to the arcuate line and the lateral walls of the true pelvis. As a check upon this procedure, further reconstructions from additional copies show that although there may be minor degrees of argument about the precise curvature of that anatomical feature, the differences produced in individual measurements of the pelvis as a result of "acceptable" reconstructions are very small indeed and do not affect the results. Even when reconstructions are supplied that would probably be at the limit of acceptability for most anatomists, the deviations introduced in the fossil as compared with those of extant forms are not large and do not affect the qualitative nature of the results. Finally, even if we make reconstructions that are frankly biased (i.e., one reconstruction has been made that aims the appearance of the fossil fragment deliberately towards man, another that turns its shape towards the apes), the form of the fossil is different enough in nonreconstructed features from both man and apes that the degrees of similarity and difference are not markedly distorted.

However, the fact that fossil remains were available from the beginning in the case of the pelvis means that this study can be somewhat reoriented as compared with that on the shoulder. Thus instead of the biomechanical estimates being based upon the entire range of primate forms as was attempted for the shoulder, the primary comparison made here is of biomechanical functions in bipedality as compared with those in generally quadrupedal locomotor modes of different types.

It is not always easy to assess whether or not osteological features of the primate pelvis are mechanically associated with the biomechanics of the pelvic girdle consequent upon variations in locomotion. Thus the shape of the obturator foramen, a character that differs widely in man from its condition in other primates cannot be obviously associated with locomotor forces. The significance of other differences can be more readily appraised. For example, the caudal extension of the ischial part of the bone, typical of quadrupedal animals, the locomotion of which is characterized by powerful and sometimes quite rapid cranio-caudal movements of the limb, gives the attached hamstring muscles a greater mechanical advantage in this movement. As retractors of the hip in the quadrupedal position, these muscles are thus better able to contribute to the power stroke in locomotion. This is most obvious of all in those quadrupedal primates that are specialized for cursorial terrestrial activity, as is the hussar monkey. This finding compares with the anatomical situation in man where, because of his upright position, the ischial tuberosity is less elongated and the attached hamstring muscles are not the only, or perhaps not even the primary, propulsive muscles during human jogging and running.

Similar variations affect the muscular architecture of the pelvic region. Many muscular features are so complex that it is rather difficult to estimate on morphological grounds what may be their mechanical relevance to locomotion if any. This is especially so of the large number of meristic variations that have been reported by comparative anatomists for over two centuries. But other features—for instance, the relative sizes, degree of pinnateness, and complexity of different functional muscle blocks—clearly can be related to some of the locomotor demands in the different forms, although it is also clear that definitive information can only come from direct studies (e.g., of the electrical activity of muscles and of tension in tendons).

It is in the single species that is habitually and uniquely bipedal, *Homo sapiens,* that a conspicuously different pattern of muscular form and disposition exists, contrasting, in fact, with all other mammalian groups. For instance, gluteus maximus muscle is exceptionally strongly developed and oriented so as to retract (extend) the hip powerfully rather than provide a relatively weak contribution to retraction and/or abduction as is the case in many quadrupedal mammals. Again, the lesser gluteal muscles (gluteus medius and gluteus minimus) in man pass lateral to the hip joint and therefore act as abductors. But in many other mammals, including nonhuman primates, they pass dorsal and caudal to the hip joint and thus, by medial rotation (or perhaps the ability to counter lateral rotation), they are important elements of retraction of the hindlimb and contribute to the locomotor power stroke. Again electro-physiological confirmation of these inferences would be most valuable.

Correlated with these and other features of the musculature of the human hip are corresponding features of the pelvic girdle and especially the innominate bone that are also strikingly unique. These can also be related to the mechanism of man's upright posture and gait. One of these certainly includes the twist of the iliac blade that brings the greater and lesser glutei into a different alignment and hence provides a different action of these muscles upon the hip joint; another may include the possession of a conspicuous and stoutly built anterior inferior iliac spine which has been associated with the attachment of the powerful ilio-femoral ligament and the origin of the large rectus femoris muscle, features apparently of the human type of bipedality.

In addition to a variety of such characters as can rather readily be correlated with biomechanics through muscular form and proportions, other features also distinguish man, and these can be readily related directly to the mechanism of load-bearing during bipedalism as compared with that in quadrupedalism. Thus, in the upright posture, the relative approximation of the sacral and acetabular articular surfaces is associated with a reduced length of bone through which the compressive forces related to the upright position may be transmitted. Again, a rotation of the sacral articular surface associated with a reorientation of the sacrum as a whole relative to the vertical axis of the pelvis when in an upright position seems also to be related to the transmission of weight-bearing forces in line with the trunk, as in man, rather than at right angles to it, as in all quadrupeds.

Thus we can see that there is little doubt about the mechanical significance of a number of features of the primate pelvis, even if all the details have not yet been worked out. Other features, partly because of the complexities of mechanical adaptation in bone form, are more susceptible to varying interpretation. Still others cannot readily be associated with locomotor function at all. Compared with the shoulder, the picture is more complicated in the pelvic girdle, because in addition to the forces impressed upon the region by contraction of the associated blocks of muscle, other sets of forces act upon the region during locomotion because of direct transmission of loads through articulating bones (a mode of loading that plays a relatively smaller part in the more mobile shoulder girdle that is almost entirely suspended by muscles). Yet other complications presumably result from visceral (for example, reproductive) functions—again a feature that is absent in the consideration of the shoulder girdle—and that, for the pelvis, cannot be estimated or allowed for in our present state of understanding.

Contrasts in Pelvic Musculature

However, notwithstanding all of these problems, the study by dissection of 167 primate carcasses, representing eight genera from the Prosimii and twenty-five genera from the Anthropoidea, has provided enough information (a) for the primary muscular arrangements to be clarified and (b) to allow the choice of a series of osteometric characters (Zuckerman, Ashton, Flinn, Oxnard, and Spence 1973). The main findings are the following. In man the four functional muscle units (extensors, flexors, abductors, adductors) are of approximately similar size, forming, as it were, four guy ropes around a single column. In contrast, in most of the nonhuman primates, the main bulk is centered around the extensors (especially) and the flexors, the adductors being considerably smaller and the abductors being extremely small (only one-tenth of the size of the extensors). In the prosimians and most anthropoids the bulk of the musculature is thus deployed for moving the limb in the craniocaudal direction with considerably less emphasis on the side-to-side balancing movements that are the perquisite of man. But it is also true that in a few New World monkeys and in the apes, medio-lateral movement seems to be somewhat more important in climbing and hindlimb acrobatic movements, and the medial and lateral musculature is correspondingly developed.

More detailed facets of the musculature include contrasts in muscular orientation, and relate principally to the gluteal muscles, on the one hand, and to the abductors, on the other. In nonhuman primates, the lesser gluteal muscles are disposed dorsally and contribute to the power stroke by medial rotation, by countering lateral rotation and possibly by retraction, though this needs confirmation by electromyography, since the disposition of the fibers is not clearly such as to make retraction necessarily an important element in the action of these muscles. In man they pass lateral to the hip joint and it is clear that they are abductors, their precise position of attachment to the greater trochanter being important in this regard. This has long been confirmed both surgically and electromyographically (see Basmajian 1972 for review).

Again, in the nonhuman primates the cranial part of gluteus maximus muscle (equivalent to the anterior portion inserting high on the femur as described by Stern 1971, 1972) passes mainly lateral to the hip joint and is a small, less powerful abductor. Only the caudal part and associated muscular slips then act as retractors. But in man the whole muscle, although homologous principally with the cranial part of the muscle of nonhuman primates, assumes a massive retractor function, not so much during walking but during raising the body at the flexed hip and, of course, during jogging, running, and jumping.

Further, the adductors (with the exception of the hamstring part of the abductor magnus, which throughout the primates is a retractor), which in man subserve mainly an adductor function, contribute significantly in the nonhuman primates to both protraction (ventral elements) and retraction (dorsal elements). These general descriptions are not at variance with the findings of Stern (1971) and of Uhlmann (1968).

Contrasts in proportional composition of pelvic muscle blocks are derived from the relative weights of each muscle and functional muscle block. Although in the case of each functional muscle group minor contrasts exist among the different prosimians, monkeys, and apes, the chief divergence is between man and the nonhuman primates. This contrast applies to the retractors and abductors. In nonhuman primates the retractors contribute some 15 percent to the total bulk of the hip musculature than is the case in man, despite the great increase of the human gluteus maximus muscle. In nonhuman primates the abductors contribute some 10 percent less to the total hip muscle bulk than in man. The total relative bulk of the protractors scarcely differs between the two groups, while the adductors vary by only some 5 percent. The short muscles account for less than 5 percent of the total in both man and the nonhuman primates.

While we recognize that relative muscle weight is at best only a crude measure of relative size of muscle blocks, and while we are certainly cognizant of the better data that may be obtained if information is introduced about degrees of pinnateness, about differences in motor unit structure, about differences in percentages of fiber types, and even about differences in physiological constants of different muscles—all of which are important in modifying the biomechanical meaning of the relative muscle weight—we recognize that to do these kinds of studies for even a pair of muscles in a single species is a big research task. To perform such studies for twenty-three or more hip muscles in over a hundred specimens is a virtual impossibility within any single investigation by one laboratory. Useful though this information may be, and much as though we would like to have it, it is not possible for it to be provided.

The Osteometric Study of the Pelvis

On the basis of these prior studies of the musculature, it seems possible to arrive at definitions of osteological features that relate at least in large part to the biomechanical demands of locomotion and posture. Thus, first, a series of characters are defined that seem to relate to the pattern of weight-bearing. As weight is transmitted through the pelvis to the leg via the sacroiliac articulation and the hip joint, it seems as though considerable mechanical relevance is associated with the positions of these two articulations both in relation to each other in the cranio-caudal and dorso-ventral directions, and in relation to the boundaries of the innominate bone as a whole at its cranio-caudal and dorso-ventral limits.

Secondly, a series of characters are defined that seem to relate to muscular pulls. These are obtained from the contrasts in relative proportions and orientations of different muscles and are approximate measures of the position of the attachment of certain muscular blocks and the length and disposition of the bony lever arms that relate these attachments to the joints. We are, of course, making simplifications here: it is not possible to measure a true lever arm because this requires definition of the axis of movement of the joint. As such axes of movement do themselves move during movement, there cannot be any single measurement of a bony lever arm applicable at all phases of a movement. But measurements based on muscular attachments and joint surfaces contain the major part of biomechanical lever lengths. Such approximations are all we can use at this time.

That all of these measurements are confined to definitions limited to a single innominate bone is demanded by the nature of the fossil material with which comparison is to be attempted.

The dimensions are projected into standard anatomical planes rather than being made directly in order to take account of the different directions in space in which such features really lie. Thus they are transposed into a set of coordinates that are orthogonal, and this makes sense when we are attempting to think in biomechanical terms. As a result, differences in angulations between features as well as differences in dimensions are also included in the data. Obviously, other orthogonal coordinate systems may be chosen, but the standard anatomical ones are readily available; and in any case, the data provide the same result as long as the coordinate axes are, in fact, in fixed relationship to one another. It is only if coordinates are not similarly fixed in all specimens that the overall form of the data changes from one specimen to the next.

However, when we take standard anthropological measurements in a variety of different directions, we usually have quite different nonorthogonal coordinates in the different species. Clearly, if the angular information is missing, such standard dimensions are inadequate for biomechanical comparisons where the size or shape of a feature is not the only information that is required. Position within a coordinate space may be every bit as important, and in many cases far more important for biomechanics than simply measurement. The size and shape of the attachment of a muscle, for example, can vary enormously without any major effect upon the ability of that muscle to act, shall we say, as an extensor of a joint. But if that attachment is twisted in its position so that it now lies, say, laterally to the relevant joint rather than dorsally, as in some other species, then that muscle becomes an abductor rather than an extensor as in the other species. Positional information obtained from a coordinate system may thus well be far more important than that obtained from direct measurements which often omit positional factors.

These facts have bedeviled many previous assessments of the Sterkfontein innominate bone, in which the resemblance in the shape of the iliac blade between man and the fossil has overshadowed the complete difference in position of that blade within the pelvis. This can easily be seen in the original descriptions that ignore the angulation problem. Thus even in the drawings of Le Gros Clark (1967) in which the innominate blades are compared, the positional differences in relation to a number of features (e.g., the acetabulum, and the ischiopubic rami) are just visible.

This is highlighted by Oxnard (1972b. 1973a), who shows pictures of such innominate bones photographed in their entirety from a number of different angles. Thus, although there is no doubt of the similarity in shape of the iliac blades of man and Sterkfontein pelvis (as can be seen from plate 1, in which the bones are photographed in the plane of the iliac blade), it is also clear that this blade is positioned quite differently in man and the fossil. In the posterior (dorsal) view this difference is evident when we inspect the picture carefully; but to the casual glance the difference is hidden because the differently twisted parts are mostly out of sight. If, however, we choose to photograph these bones at exactly the same angle but from the anterior (ventral) surface, it then becomes extremely clear just how the two specimens of man differ from those of the ape and the fossil. In the former the ischiopubic part of the bone still points to the left; in the latter it has become twisted so far around that it points clearly to the right (plate 2). Plate 3 shows the angular position of the iliac blade in relation to the plane of the ischio-pubic ramus. Their relative orientation differs between man and the fossils, and, again, it is with an ape that the angular relationship in the fossil is most consonant. Although an innominate bone from a particular ape, the chimpanzee, has been chosen for these comparisons, similar information results if we use as the nonhuman specimen one from a gorilla or an orangutan. Were we to use one from a gibbon or any monkey, yet other marked positional differences would be immediately apparent.

That this change in position of these different parts of the bone would undoubtedly have importance in considering locomotion can be seen from figures 28 and 30. Here an attempt has been made to put lateral views of these elements in man, *Australopithecus,* and chimpanzee into positions that might relate to both bipedality and quadrupedality in all species. In figure 28 bipedality is simulated and we can rather clearly see the similarities between the fossil and the ape even although we cannot be certain what is the correct orientation for the fossil pelvis upon the femur (fig. 29 provides three possibilities—two extreme and one central). In figure 30 the same three pelves are simulated in a typical quadrupedal position; again there can be little doubt as to the similarity of the fossil simulation with that of the ape. Here several views of the fossil are not required because, from the form of the bone, we can be fairly certain of its orientation in such a simulation. In both of these cases we have been rather careful to avoid frontal or oblique views, in which the shape similarities of the iliac blade of the fossil to that of man overpower the positional differences between the fossil and man.

The actual dimensions that we have taken on each of 430 pelvic girdles of forty-one genera of primates are detailed in the paper by Zuckerman, Ashton, Flinn, Oxnard, and Spence (1973), but in essence they comprise measures of nine features of the innominate bone. Four dimensions relate to joint positions: (a) the dorso-

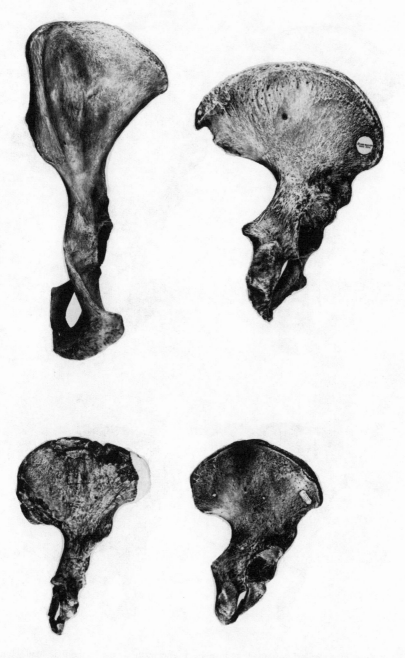

Plate 1.   Photographs of the Wenner-Gren casts showing the similarity in shape of the iliac blade in man and the fossil. Upper left, chimpanzee; upper right, human female; lower left, *Australopithecus;* lower right, specimen from a pygmy that happens to be approximately the same size as *Australopithecus.* Dissimilarity between man and the fossil in the lower part of the bone is not clearly evident because it lies behind the rest of the bone.

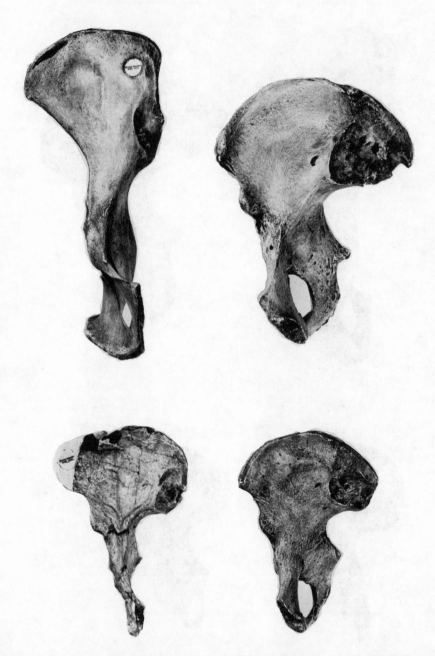

Plate 2.   Photographs similarly oriented, as in plate 1, and in the plane of the iliac blade, but viewed from the opposite side. The similarity of the shapes of the iliac blades of the human specimen and *Australopithecus* are not in doubt, but now we can clearly see the similarity of the lower part of the bone in the ape and *Australopithecus,* in which they are both completely different from the men. In the human specimens the pubic ramus points to the left of the picture, in the ape and *Australopithecus* it points to the right.

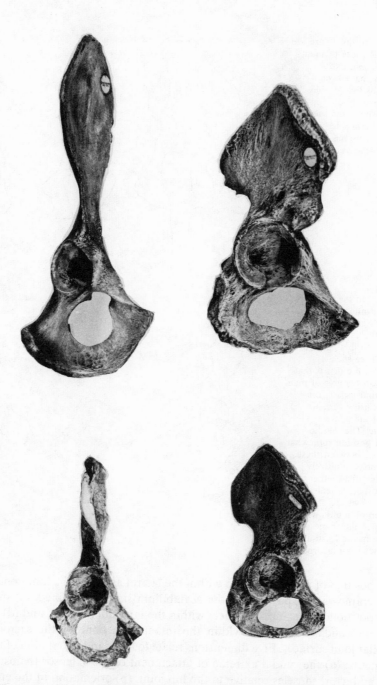

Plate 3.   The relationships of the two previous pictures are further outlined when we compare the same four casts viewed in the plane of the ischio-pubic part of the bone. Here the similarity of ape and *Australopithecus* is most marked; both are clearly different from the men; the difference obviously rests in the relative orientations of the different parts of the bone.

Figure 28.   In this figure the pelves and femora of man, *Australopithecus*, and ape, respectively, have been orientated as though they were customarily used in the bipedal position and viewed from the lateral side. Similarities between the ape and *Australopithecus* are clear.

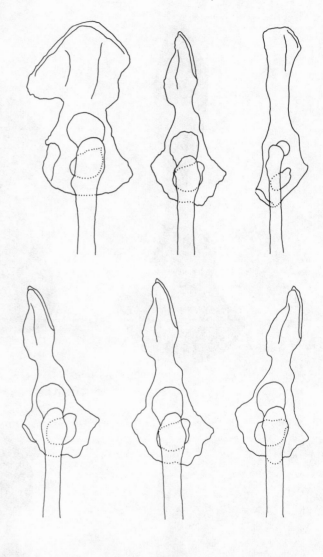

Figure 29.   In the previous figure we cannot be certain what the orientation of the australopithecine specimen should be. In this figure are provided three such orientations; the first is that using certain features of man as being appropriate, the last is that using certain features of the ape as appropriate. The middle figure is a position somewhat intermediate between these two that looks intuitively more correct than either. It is most important to realize that these views are all true laterals—no obliquity, which would reveal the shape of the iliac blade in the fossil and the ape, has been incorporated.

ventral position of the acetabulum within the lateral aspect of the innominate bone, (b) the cranio-caudal position of the acetabulum similarly defined, (c) the dorso-ventral position of the auricular facet within the innominate bone, and (d) the relative cranio-caudal separation within the innominate bone of the sacroiliac and acetabular joint surfaces. Five dimensions relate to locus and orientation of muscular attachments: (a) the caudal extreme of attachment of the extensor (hamstring and certain adductor) muscles relative to the hip joint, (b) orientation of the entire iliac blade (giving origin on part of its dorsal edge to gluteus maximus in man and superiorly to many abdominal muscles) relative to the position of the hip joint, (c) orien-

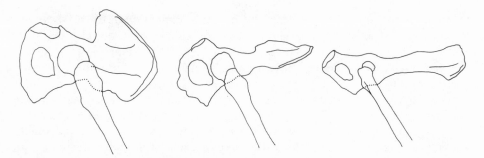

Figure 30.   The pelves and femora of man, *Australopithecus,* and an ape have been placed as
though all three were quadrupedal. The similarities of the ape and *Australopithecus* are most evident.
And in this case the point is so obvious that there is no need for additional simulations of the fossil.

tation of the ventral part of the iliac blade alone (the part giving origin on the dorsal
surface to the lesser glutei and on the ventral surface to the iliac muscle) relative
to the hip joint, (d) the dorso-ventral position of the anterior superior iliac spine
(possibly related to the attachment of the powerful ilio-femoral ligament and rectus
femoris muscle of man) relative to the innominate bone as a whole, and finally, (e)
the dorso-ventral position of the anterior superior iliac spine relative to the position
of the acetabulum.

These variables are taken from all of the specimens available to us, but, as marked
variation in size exists, not only between individuals of the two sexes and individuals
within a genus but especially between different primate genera, it seems necessary
to try to compensate in some way. This is done by standardizing each dimension
against a quantity that can be taken as indicative of the general size of the animal
and of its complete pelvic girdle. The dorso-ventral dimension of the pelvis is available for every specimen, is a major dimension of the innominate bone that reflects
its overall size, and also appears to be unrelated to any of the biomechanical features
previously outlined. It seems in some ways analogous to the cross-sectional measurement of the vertebral canal that has also been used for standardization by a number
of investigators in evolutionary studies at higher taxonomic levels (Radinsky 1967;
Wanner 1971). An analysis of generic information for the primates shows that this
dimension correlates fairly closely ($r = 0.9$) with the crown-rump dimensions obtained, in the case of man, from Bean (1922); in that of most nonhuman primate
genera, from Elliot (1913) and from Napier and Napier (1967); in that of the
orangutan, from Schultz (1941); and for those of the chimpanzee and gorilla, from
field records made available to us by the Curator of the Powell-Cotton Museum
(Birchington). Of course, it would have been preferable to have standardized by the
actual crown-rump length, a genuine external measure, of each specimen. But it is
only for the chimpanzee and gorilla that this is possible. Accordingly we have used
the relationship at the pooled level between the crown-rump length and the dorso-
ventral measure of the pelvis to confirm that we have not gone far wrong in using

this internal measure. The projected dorso-ventral dimension of the pelvis appears to reflect fairly, not only the general size of the pelvic girdle, but also that of the animal as a whole, although, as pointed out by Biegert and Maurer (1972), the overall size of the limbs is not necessarily constantly related to the size of the trunk within a series of different primate genera. This is probably the best that we can do.

Size is then compensated for by regression adjustment. First, each linear, although not angular, measure is expressed as an index using the dorso-ventral measure of the bone as the denominator. Second, for each dimension, both linear and angular, a regression is calculated, the dorso-ventral dimension of the pelvis being the independent variate. After logarithmic transformation of the data (necessary to eliminate curvature in the regression lines indicative of the allometric relationship between each functional dimension of the pelvis and overall size as reflected by the dorso-ventral measurement), the regression lines for each genus are found, within the limits of sampling error, to be similar enough across the primates to allow an overall regression to be calculated. The slope of this varies from dimension to dimension across the primates, thus indicating a different allometric relationship between each dimension and the measure of overall size. These slopes are then used to apply correction factors to the value of each dimension in each specimen of each genus in order to assess what would have been their values if all the specimens were of the same size. This mode of regression adjustment is standard. However, in subsequent analyses, caution has dictated to us that studies be carried out on both regression-adjusted and nonregression-adjusted data. It turns out that the two sets of results obtained are generally parallel. But because separations are somewhat better with the regression adjusted data, it is these latter that are reported on in all definitive studies.

Results of Univariate Studies

Each individual variable is examined using univariate statistics such as the mean, variance, and standard error, and these show that the principal pattern of contrasts throughout the primates varies quite remarkably. In the case of some dimensions— for example the cranio-caudal position of the acetabulum—the variation characteristic of the nonhuman primates as a whole completely overlaps that characteristic of man (fig. 31). In the case of other dimensions, however, for instance the position of the anterior superior iliac spine relative to the center of the acetabulum, the range of variation from man is totally outside that representing the remainder of the primates (fig. 32).

The position of man is also important in relation to that of the different apes. Man does not overlap with the great apes, and even in the case of the lesser apes, overlap is found in only a single dimension (the cranio-caudal position of the hip extensors relative to the hip joint, fig. 33). Paradoxically, therefore, as judged from univariate examination of these particular biometric parameters of the pelvis, man may *seem* rather more distantly separated from his nearest genetic relatives, the great apes, than he is from the other nonhuman primates as a group.

Although interest is naturally aroused by consideration of these many comparisons among the primates as a whole, it is upon the resemblance and differences

Figure 31. Univariate
analysis of the cranio-caudal
position of the acetabulum
measured as the relative
increase of length between
them. Means and 90 percent
limits are provided for
hominoids. Extreme means
and 90 percent limits only
are given for all nonhominoid
primates. Three positions are
provided for three slightly
different reconstructions of
the fossil.

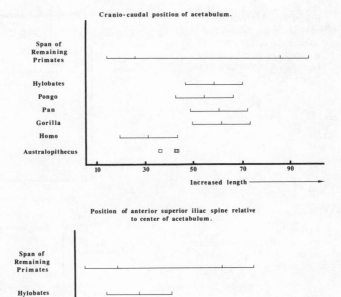

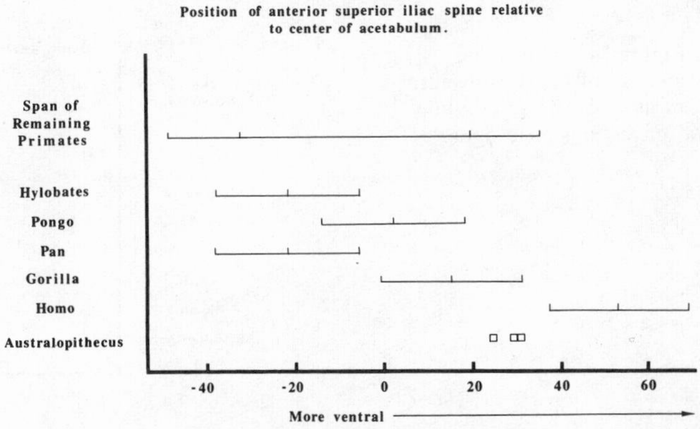

Figure 32. Univariate
analysis of the position of the
anterior superior iliac spine
relative to the center of the
acetabulum. Conventions as
in figure 31.

of the Sterkfontein pelvis to man and the apes that most attention focuses. Here the
univariate pattern of results (provided by fig. 31 through 36) is the following: *Australopithecus* generally falls in a variable position, resembling man in certain instances, the great apes in others, and being intermediate in yet others. In particular,
if we look at the three examples of the dimensions relating to the relative positions
of the sacroiliac and hip joint, the fossil shows similarities with man in the dorso-
ventral position of the acetabulum (fig. 34), with both man and an ape in the dorso-
ventral position of the auricular facet (fig. 35), and with man and a variety of other
nonhominoid genera in the cranio-caudal position of the acetabulum (fig. 31). In
these features in toto, however, the fossil seems to resemble man rather more closely
than the great apes.

In the case of those examples of dimensions relating to muscular disposition and
leverages, *Australopithecus* again resembles many different nonhuman primates
and usually also one or other of the great apes (fig. 32, 33, and 36). The fossil usually
differs markedly from man, although in the case of two dimensions it is, though just
outside the 90 percent fiducial limits for man, nevertheless rather closer to man
than to the great apes (e.g., fig. 36). For the five dimensions as a whole, however, it
seems as though the fossil resembles rather more the apes than man.

Figure 33.   Univariate
analysis of the position of
attachment of the extensor
musculature relative to the
hip joint. Conventions as in
figure 31.

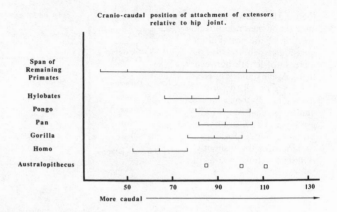

Figure 34.   Univariate
analysis of the dorso-ventral
position of the acetabulum.
Conventions as in figure 31.

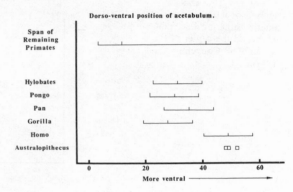

Figure 35.   Univariate
analysis of the dorso-ventral
position of the auricular
facet. Conventions as in
figure 31.

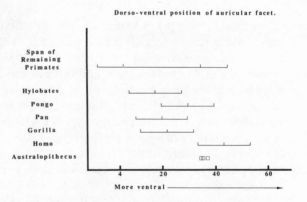

But on the basis of the total of nine such univariate examinations it is really not
possible to say which species the fossil most resembles. It is clear that a question of
this type can be answered only through the use of a multivariate statistic which can
give a precise quantitative assessment of the degrees of difference and resemblance.

Figure 36.   Univariate
analysis of orientation of iliac
blade ventral to the dorsal
buttress of the innominate
bone. Conventions as in
figure 31.

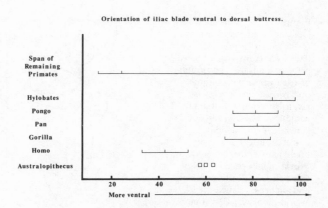

The Multivariate Analysis

A first way of viewing the multivariate results can be obtained from a bivariate plot of the first two canonical axes for all primates (fig. 37). This shows that the various primates are strung out mainly along the first canonical axis with prosimian and most nonprehensile-tailed New World forms primarily at the right-hand end of this axis, and Old World monkeys, *Cebus* and prehensile-tailed cebids, and lesser apes mainly at the left-hand end. The great apes are arranged in linear system at right angles to this, and it appears as though man is somewhat close to the most distal ape, which happens to be the gorilla. He is also almost as close to several other nonhominoid primates. The Sterkfontein fossil (three reconstructions) appears nearly midway between the gorilla and man, although slightly nearer to the ape.

However, in an analysis as complex and multifaceted as this, it is clear that there is much information that is contained within canonical axes higher than these first two. Although we can examine each higher canonical axis one by one or in pairs, it appears as though a good way to display further the high-dimensional information is through the minimum spanning tree superimposed upon the bivariate plot of the first two canonical axes. This is shown in the next two figures, 38 and 39.

Here, the various primates are arranged in such a way that a number of interesting conjunctions of genera occur. First, those of New World monkeys and prosimians which are generally quadrupedal are considerably separated from the generally quadrupedal Old World monkeys. Emanating from the block of New World and prosimian quadrupeds are "side chains" that consist of, respectively, the vertical clinging and leaping genera, *Galago* and *Tarsius,* the vertical clinging and leaping indrids, the slow-clinging lorisines, the aberrant individual genus *Daubentonia,* and finally the (probably nonprimate) genus *Tupaia* (fig. 38). Secondly, the Old World quadrupedal species that also fall together in a nucleus have likewise a series of side chains that proceed from them. One of these contains *Papio* and *Mandrillus,* and another *Erythrocebus,* the most terrestrial of these species. Yet another contains the various colobines (e.g., *Nasalis*). A fourth comprises a set of connections that encompasses many of the more acrobatic species ranging from *Cebus* through the

Figure 37.   Canonical analysis of nine dimensions of the pelvis in primates. The data were regression-adjusted prior to analysis. The position of each genus within the first two canonical axes is given by a dot, and the contours around groups of genera are one standard deviation in extent. A = prosimians and most New World monkeys; B = lesser apes, all Old World monkeys, and prehensile-tailed New World monkeys; C = man; D = great apes; X = positions of three reconstructions of *Australopithecus*.

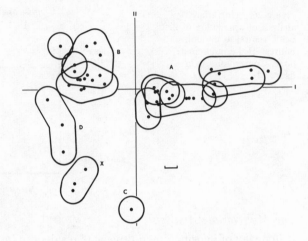

Figure 38.   Because much of the information in the analysis of figure 37 is found in canonical axes higher than the first two, an attempt is made here to provide the additional result. The directions given by the first two canonical axes are used as a base upon which to hang the true connectivity and true distances provided by the minimum spanning tree derived from the complete matrix of generalized distances of the same data (compare figure 25, chapter 2). Because of the complexity of these higher dimensions, this figure is confined to the analysis of those groups on the right-hand side of figure 37—the prosimians and certain New World monkeys. The star-shaped arrangement of genera is clearly evident. Those centrally located are basically quadrupedal; the various rays or arms emanating from the central nucleus contain genera that practice locomotor patterns including other modes of movement such as leaping and clinging. Unconnected

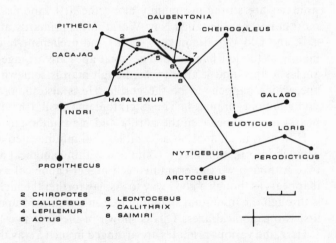

2 CHIROPOTES	6 LEONTOCEBUS
3 CALLICEBUS	7 CALLITHRIX
4 LEPILEMUR	8 SAIMIRI
5 AOTUS	

genera in the diagram are not (and cannot be) correctly placed in relation to one another. Connections that are dotted are near minimum, and, though they are of incorrect length in this diagram, they are those particular relationships that happen to be close to the minimum ones and that might quite easily represent the true minimum distances should error be better controlled, samples be larger,

and so on. It is noteworthy that there are many such near minimum connections within the central nucleus, thus indicating that these forms are all similarly placed; there are no such connections between different peripheral arms of the star, indicating that these arms are truly separated and are only connected to one another through the central nucleus.

Figure 39. This figure
provides the equivalent result
as figure 38 for the other half
of the canonical analysis of
figure 37. Conventions and
discussion as in figure 38.
In this case, however, the
central genera are those Old
World monkeys together with
capuchins (*Cebus*), which
are basically quadrupedal.
The various rays or arms
emanating from the nucleus
are those genera that practice
locomotor patterns that
include elements markedly
different from the most
obviously quadrupedal forms.

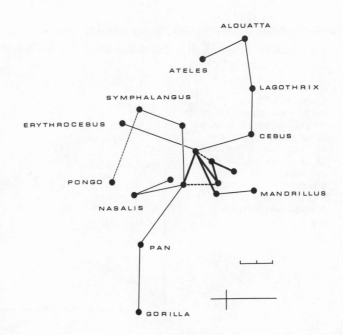

prehensile-tailed New World genera; a fifth comprises Asiatic apes, and a sixth the African apes (fig. 39). There are, however, some near minimum connections (not shown) that ally these last three arms together in a complex, twisted set of connections.

Man does, of course, have an extant genus to which he is closest, and that is the gorilla. However, man is almost equally close to the orangutan, to the proboscis monkey, to the woolly monkey, and to the indri, and it is clear, therefore, that there is no special meaning to the fact that the distance from the gorilla is actually absolutely the smallest. All of these distances are large (*circa* twelve standard deviation units) compared to those among almost all other genera; moreover, the differences among these large distances are less than two standard deviation units and thus could well be due to sampling errors, chance, or possibly other unknown factors. In other words, man is uniquely different from all other living species in the features examined here.

When we also examine the position of *Australopithecus* in this matrix, it seems clear that it falls roughly intermediately between man and the gorilla, possibly slightly closer to the latter. It is certainly not the same as man, being 7.6 standard deviation units distant from him; likewise, however, it is clearly distinct from the gorilla by 7.5 units.

As far as this analysis is concerned, it is the multivariate study that establishes that the relationships of the fossil are with man and the apes and not with the other forms. Accordingly, a second analysis of the nine dimensions is carried out in which the great apes and man only are compared. The results confirm the picture already presented in figure 37. It appears as though the fossil is intermediate between man and the apes (fig. 40).

Figure 40. Bivariate plot of the first two canonical axes of the study of nine dimensions of the pelvis. The three reconstructions of the fossil (A, B, C) fall roughly intermediate between the great apes (1, 2, 3) and man (4). The marker provides two standard deviation units.

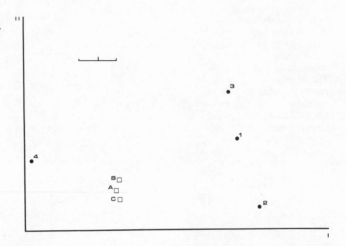

It is also possible to make two further analyses, one of each subset of the dimensions. Perusal, for instance, of the first two canonical axes for the analysis of the four dimensions related to joint position shows that the fossil is remarkably similar to man (fig. 41). Conversely, however, the equivalent examination of the five dimensions relating to muscular attachment and disposition, provides a picture in which the fossil is rather more similar to the apes, and specifically to the gorilla (fig. 42).

Bearing in mind the warnings of an earlier chapter, we may wonder if there are any other differences in any of these studies that are contained within yet higher canonical axes. In fact, we already know that there are, because of the differences between the canonical analysis of all of the primate genera (fig. 37) and the minimum spanning tree derived from the matrix of generalized distances and superimposed upon the directions presented by the canonical plot (fig. 38 and 39). An examination of the higher axes from the study of apes, man, and *Australopithecus* alone shows that the picture derived from the first two canonical axes is not the whole story.

Let us therefore examine this set of results using the sine-cosine multidimensional display of Andrews (1972, 1973). Figure 43 shows the result of the four dimensions that relate to the relative positions of the joints within the pelvis. This demonstrates that the prediction of the first two canonical variates—that the fossil markedly resembles man—is indeed the picture also provided when we include all the information in the higher axes. Likewise, figure 44 provides the result for the five dimensions which are measures relating to muscular positions and attachments. Again this confirms only what we already know from the examination of the first two canonical axes of that study, namely that the fossil markedly resembles the apes as a group, although not especially any single ape.

Figure 41.   Bivariate plot of
the first two canonical axes
of the study of four
dimensions of the pelvis
relating to the positions of
joint surfaces. The three
reconstructions of the fossil
(A, B, C) fall very close to
man (4) and well away from
the great apes (1, 2, 3).

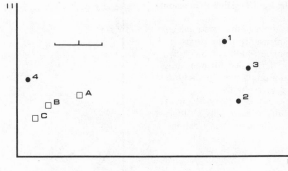

Figure 42.   Bivariate plot of
the first two canonical axes of
the study of five dimensions
of the pelvis relating to the
positions of muscle
attachments. The three
reconstructions of the fossil
(A, B, C) lie close to the
great apes, especially the
gorilla (3) and the
orangutan (1).

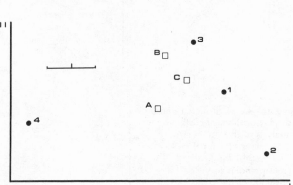

However, when we examine the total set of canonical variates all taken together
for the total of nine dimensions combined, figure 45 reveals that the picture is not as
simple as saying that the fossil is intermediate between man and apes. If we follow
the plot from left to right, we can see that the group of apes forms a rather tightly
arranged set of curves that, in the diagram, are shown only by their conjoint envelope. We see also that the curve for man is totally different from the envelope for the
apes, so that we can confirm that man is markedly outlying from the apes. But if we
examine the curve for the fossil, we note that, although it lies rather neatly between
the curve for man and the envelope for the apes at the left-hand side of the plot,
it does not follow this position consistently across the plot. At the right-hand end the
fossil is outlying to the curve for man and the envelope of the apes. This means that
the fossil is not a simple intermediate; rather, the relationship is triangular: man is,
of course, uniquely different from the apes, but *the fossil is uniquely different from
each.*

Conclusions for the Australopithecines

What may be the meaning of a result like this? The contrast in the features of the
pelvis of the nonhuman extant primates with that of man appears to correlate with
the primary difference between generally quadrupedal function in nonhuman
primates and the bipedalism of man. Although many nonhuman primates can and

Figure 43. High-dimensional
display of the canonical
analysis of four articular
dimensions of the pelvis. The
plot for the fossil is closely
similar to that for man thus
confirming that the picture
provided by the two-
dimensional plot of figure 41
contains essentially all the
discriminating information.

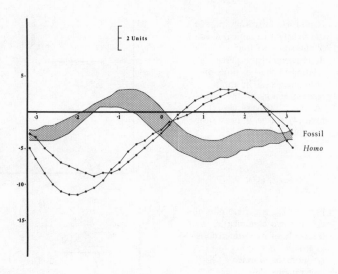

Figure 44. High-dimensional
display of the canonical
analysis of five muscular
dimensions of the pelvis.
The plot for the fossil is
closely similar to the
envelope of the great apes
thus confirming the two-
dimensional picture provided
by figure 42.

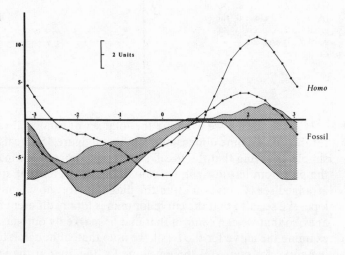

do walk bipedally, the manner in which this is done is clearly nonhuman and presumably the structures that allow it are conditioned by variations on the quadrupedal theme. Thus the mere fact that the pelvis of *Australopithecus* is uniquely
different from that of all extant nonhuman primates suggests that it must have been
doing something uniquely different from them. Such a difference was probably a
form of bipedalism.

But the unique difference between the fossil and man must also mean something;
whatever that form of bipedalism was, it must have differed uniquely from that
characteristic of man today.

Is there more to be learned? Perhaps we are being provided with some view of the
nature of the adaptation in the fossil. The fact that the arrangement of the joints is
the feature resembling man suggests that the fossil may well have been habitually

Figure 45. High-dimensional
display of the canonical
analysis of nine dimensions
of the pelvis. The two-
dimensional plot (figure 40)
suggests that the fossil is
intermediate between man
and the apes. The higher
dimensional plot shows that
this is not so. For the left-
hand part of the plot the
curve for the fossil is
intermediate; but at the
right-hand end it is man that
is intermediate. This means
that the relationship of man,
fossil, and apes is truly
triangular; the fossil is
uniquely different from each
and not intermediate when
we take all the information
into account.

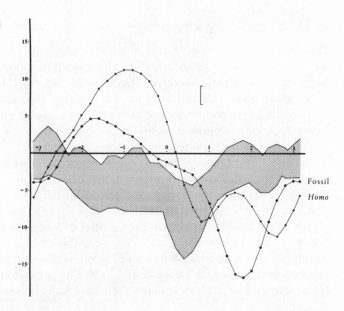

upright with an arrangement of joints that allowed it to support its weight in line
with the trunk as in an upright creature rather than at right angles to it as in four-
footed forms. But the fact that the arrangement of some of the muscular attachments
is really rather more like those of the apes than man suggests, perhaps, that whatever
the nature of the evolutionary adaptation upon which *Australopithecus* is embarked,
the function of the musculature is essentially like that in a number of nonhuman
forms. This must mean that the biomechanical nature of the bipedalism of the fossil,
if it existed, must have been markedly different from that of modern man. It is likely
that, like the chimpanzee, gorilla, and other nonhuman primates, the fossil, when it
moved upright, may have been better adapted for running rather than for a striding
gait. It is also likely that this creature may still have been able to do the things that
nonhuman primates can do well—namely, climb trees acrobatically and move
quadrupedally. Modern man is either inept at or totally incapable of these functions.

It is certainly clear that *Australopithecus* was not some simple morphological
intermediate between apes and man, nor was he displaying some simple intermediate form of locomotion between quadrupedalism and bipedalism. But it is also
clear—and this is much less well recognized—that the fossil is *not* like man in structure and presumably could not have been moving like man today; human bipedal
walking is *not* established in this genus on the evidence of this pelvis. We just do
not know what the animal is doing; there is some evidence for a type of habitual
bipedalism; there is some evidence for climbing and quadrupedal abilities; there
must presumably have been some functional mix that has not been determined at
this time. When now we return to the conclusions based upon the shoulder studies
(earlier in this chapter, under the heading "The Shoulder of Australopithecine
Fossils"), we may feel stronger about the speculation that the shoulder, too, suggests
a creature capable of considerable climbing activities within trees.

4

An Element of the
Primate foot—The Talus

Previous Studies on the Talus

Because we have now obtained from two rather different anatomical regions, the shoulder and the pelvis, information about the australopithecines that is somewhat different from what is generally believed to be the case, it behooves us to look again at as many other anatomical regions as we can. One such region that is capable of providing very good evidence about locomotion and posture is the foot. Perhaps nowhere among the primate is this as obvious as in man, where the form of the foot, with its longitudinal and transverse arches and its fixed big toe, seems clearly related to the demands of bipedal standing, walking, jogging, and sprinting. However, even among various nonhuman primates, differences in the form and proportions of the foot (for instance, short toes in the terrestrial baboons and patas monkeys; long digits in the arboreal spider monkeys, gibbons, and, most of all, orangutans), are clearly related to biomechanical peculiarities of locomotion. The most unique adaptations of all are found in those species capable of highly specialized hopping, for example, tarsiers and bushbabies, and perhaps to lesser degrees, dwarf lemurs and mouse lemurs, all of which are distinguished by somewhat lengthened tarsal bones (calcaneus and navicular) whereby their hindlimbs possess an additional lever arm in leaping.

The entire foot skeleton of a fossil primate would provide much information about its mode of locomotion. However, whole foot skeletons are rather rarely found. One of the best specimens is that of Olduvai Hominid 8, and even here critical parts are missing (metatarsal heads and calcaneal tuberosity). These components are especially important because they contact the ground during bipedal stance and gait, and their presence is therefore necessary for estimation of forces and lever arms. If these missing parts are reconstructed on the basis that they resemble the modern human foot, then the lever arms can be estimated as more like those of man than of other primates (Day and Napier 1964). But other bases for reconstruction would lead to different results.

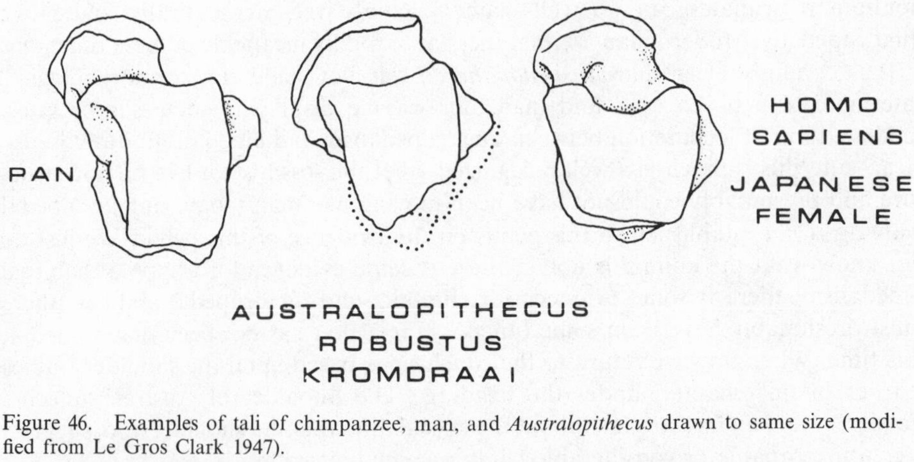

PAN

HOMO
SAPIENS

JAPANESE
FEMALE

AUSTRALOPITHECUS
ROBUSTUS
KROMDRAAI

Figure 46.   Examples of tali of chimpanzee, man, and *Australopithecus* drawn to same size (modified from Le Gros Clark 1947).

One particular element of the foot, the talus, is more frequently and completely preserved than the other bony elements (fig. 46). Lying in a central topographic and functional relationship to the other tarsal bones, its shape reflects to a large degree the structure and function of the entire foot. As a result, therefore, studies of the talus are of particular value in attempting to elucidate the evolution of function in the foot among higher primates, and a series of studies has been carried out in recent years with this goal in mind. One much cited (e.g., see Jolly 1972; Pilbeam 1972; Simons 1972) study, that of Day and Wood (1968, 1969), compares the talus of fossils from Spy, Skūhl, Kromdraai, Olduvai, Rusinga, and Songhor with those of various modern human groups and of the African great apes. The conclusions of that work relate to the African ape-human separation and suggest that the fossils from Spy and Skūhl (Neandertals) are closely similar to modern man; both, of course, are striding bipeds. The fossils, from Songhor and Rusinga (dryopithecines—*Proconsul*) are closely similar to modern African apes and this is interpreted as relating to terrestrial quadrupedalism. The specimens from Kromdraai (*Paranthropus* [=*Australopithecus*] *robustus*) and Olduvai ("*Homo habilis*" = *Australopithecus africanus*) are fairly similar to each other and are generally intermediate between the African great apes and modern man, though judged perhaps as closer to man. These similarities are derived from canonical axis one, the axis separating bipeds from quadrupeds (although also, of course, separating men from African apes). Axes above the second were not discussed in that study. This result for the australopithecines is contrary to the findings of Lisowski, who, in a univariate investigation (1967) suggests that, save for Neandertals, none of these fossils are closely similar to man.

The similarities suggested by Day and Wood do not imply taxonomic affinities. It is obvious that, on the basis of time differences alone, none of these fossils can belong to the taxonomic groups tested. Such similarities as these authors can see, they attribute to locomotor adaptation (this view is still held by Day in 1974).

More recently, Oxnard (1972c) has shown that the conclusions of Day and Wood should be modified because the authors did not take into account all of the information in their analyses. The modification shows that Neandertals (as represented by Spy and Skūhl) differ more from modern man than was previously realized, though nevertheless these groups are still close to one another. Presumably they are similar enough that they may still be claimed as bipeds, a result in accord with what we know of Neandertals. The specimens of dryopithecines (*Proconsul*—from Rusinga and Songhor), however, actually differ more from the African apes than even these latter do from modern man, a conclusion which may be incompatible with the idea that these fossils moved like the African great apes. The australopithecines (both Kromdraai and Olduvai), rather than being roughly intermediate between the African apes and modern man, are uniquely different from each, being morphologically more distinct from both modern man and African apes than these latter are from each other. Again this may mean that, in locomotor terms, these fossils are more different from the African great apes and from man than was previously supposed.

In addition, other suggestions may be drawn from these conclusions. First, the new interpretation suggests that the relationships among the fossils themselves, which could not be certainly examined in the previous study, may be of especial importance. Thus it now appears that these particular specimens of *Proconsul* are morphologically somewhat similar to those of *Australopithecus*, both being quite different from the extant African great apes, although this conclusion is only tentative (Oxnard 1972c). Presumably, therefore, there may be some locomotor similarities between these two fossil groups separated by several millions of years of evolution. Secondly, the removal of the African great apes from a position as being immediately comparable to these fossils suggests that other extant species should also be used for comparison (Oxnard 1972c). Though the importance of the gorilla and chimpanzee as the closest extant relatives of man can scarcely be challenged, osteological studies of the type discussed here relate less to genetic propinquity and rather more to biomechanical adaptation to behavioral function. Thus it is appropriate to include forms that, though genetically far distant from man and African apes, cover the spectrum of biomechanical possibilities as far as possible (i.e., many arboreal genera, of both apes and monkeys). Perhaps it is to locomotor variants similar to these nonterrestrial forms that adaptive similarities may be important. Thirdly, the more recent multivariate consideration of fossil tali (Oxnard 1972c) shows that information of considerable importance to the comparisons may be contained in higher dimensions which are sometimes not considered. Accordingly, it is appropriate to examine closely such information in the discussion of relevant differences.

The Talus—A New Osteometric Study

All these points are further investigated in a new study by Lisowski, Albrecht, and Oxnard (1974a, b, 1975), who examine another suite of linear and angular measurements of 322 tali of a total of thirteen extant primate groups and of fifteen tali (of which only two were casts) of five fossil groups. This new study attempts to take account of all aspects of the information presented. The materials used in the new study comprise series of tali from certain extant primate genera *(Ateles, Macaca, Papio, Colobus, Hylobates, Pongo, Pan, Gorilla,* and *Homo)*, a series of tali from certain fossil or subfossil genera, *Homo* (as represented by Bronze Age man from Jericho), *Proconsul* and *"Limnopithecus,"*[1] together with single fossil tali representing *Homo neanderthalensis* (from Kiik-Koba), and two representing *Australopithecus* (*"H. habilis"* and *Paranthropus robustus*). Of these specimens, only a single example of *Proconsul* and the specimen of *Paranthropus* are casts; the remainder of the materials are examined in the original.

The methodology of this study differs somewhat from that in the pelvic and scapular investigations, in that the measurements used here are less specifically chosen for biomechanical relevance. The talus is such a small bone, that, although many aspects of its morphology are undoubtedly related to biomechanical functions,

---

1. Although labelled *"Limnopithecus"* when the data were taken from these particular specimens at the Royal Free Hospital Medical School a number of years ago, these specimens cannot actually be from that genus. Further study of additional fossil specimens from Africa that are presumably not australopithecines is described by Lisowski, Albrecht, and Oxnard 1975.

it is likely that a large part of its morphology can be represented within a reasonable number of dimensions. A second consideration in the choice of measurements relates to the previous study of Day and Wood. While no attempt is made to copy exactly the measurements of Day and Wood, it is useful if approximately similar measurements comprise at least a part of the suite of dimensions. This allows some comparability of results at a qualitative level. Finally, while the selection of suitable dimensions for inclusion in a multivariate statistical study is always a critical matter, limits are once again placed upon the freedom of choice in this study because of damage to the available fossils.

The measures chosen for analysis are undoubtedly related to the functions of the complex of joints around the talus, even though they have not been chosen with this in mind. For instance, it seems clear that measures of the trochlear surface reflecting its shape must be related in part to the movements that are allowed between that surface and the tibio-fibular complex, the ankle joint proper. They undoubtedly relate to the transference of weight and movement between the leg and the foot. This is of marked importance in differentiating human walking from four-footed locomotion on the ground, and it might also be expected to differentiate between four-footed walking on on the ground on a stable substrate as compared with four-footed or four-handed grasping during arboreal locomotion. The measurements available to characterize the trochlear surface are the anterior, transverse, and posterior breadths of the trochlea together with a measure of its curvature.

Also important in this regard, and especially associated with the relatively fixed big toe of man, is the relationship between the entire body of the talus (or perhaps more specifically the trochlear surface) and the head of the bone articulating with the forefoot. These presumably relate to the mobility and prehensility or lack of it evidenced by many primate feet. The relationships can be expressed in two ways: the angulation of the head as indicated by the set of the talar neck upon the bone, and the twist of the head upon the neck when viewed along its long axis.

Finally, the overall height of the talus may well be expected to be related to the degree to which forces are acting in a vertical and straight system, as in man during bipedalism, or in an inclined and bent system, as in many apes and monkeys in grasping and holding at an angle. Two general measurements are available to express this: the maximum medial and lateral heights of the talus which when taken separately express elements of its overall shape, and which when taken together express the positional twist of the bone in the frontal plane.

The precise definitions and mode of measurement for the eight dimensions can be obtained from the original publication (Lisowski, Albrecht, and Oxnard 1974). Before passing to an examination of the univariate and multivariate results obtained from these data, a number of other matters obtrude themselves. These relate to some of the problems always associated with this type of biometric research.

Problems in the Study

Thus, for instance, we have always clearly recognized problems in morphometric data that relate to such matters as the need to pool, in the different investigations, sets of data that are not homogeneous: for example, first, sex, racial, and subspecific or specific subgroupings of genera; secondly, the spectrum of differences that may relate to the different ages of the specimens; and thirdly, problems relating to overall difference in size, not only within groups (often related, at least in part, to sex and age) but more specifically between groups (macaques are intermediate-sized primates, gorillas are very large).

In all of our own studies these problems have been recognized, and Oxnard (1973a) attempts to classify them and notes the various tests that have been applied. In the case of the shoulder studies, simple tests attempt to control for sex, racial, and subgeneric differences. Age effects are partly avoided by restricting the study to adult materials. Size is allowed for by the old method of the use of indices. It is only more recently (Oxnard 1969b) that attempts have been made (e.g., principal components analysis) to remove "size" effects statistically, and even here the procedure that is carried out is only tentative and is open to some criticism. Its best defense is, perhaps, that the effect of trying to remove size in this way produces an answer qualitatively similar to that of the simpler studies, an answer that seems to make some sense.

As the investigations on the pelvis have moved underway, our own sophistication in these matters has increased, and it is now possible for us to carry out a series of yet better procedures. We have already described the regression adjustment method for the pelvic data, something that we could not attempt in the shoulder and which, because of the way in which the shoulder data are recorded, cannot be performed on these data retrospectively.

For the talus, however, work carried out even more recently (Lisowski, Albrecht, and Oxnard 1974b) has been able to provide even further preliminary tests to strengthen the study in these respects. That we can do so stems mainly from the foresight of my collaborator in this study, Professor F. P. Lisowski, who, although he collected the data some years ago, nevertheless maintained the full data set in a manner that allows us to examine it in these different ways.

Thus, first, we can study a subset of six variables for which information about several human racial groups is of relevance. This allows us to test the effects of pooling in the one genus *Homo* where subgeneric groups are available. The data are extensive enough to allow this subsidiary study to be done for the first six dimensions taken upon the human subgroups—mixed,[2] Jericho, Australian, Vedda, and Andamanese man, together with the nonhuman genera: *Gorilla, Pan, Pongo, Colobus, Papio, Macaca,* and *Ateles.* This procedure allows us to assess the degree of distortion that may arise if a known complex assortment such as *Homo* is treated as a simple group. This procedure should be carried out for all genera (especially, for instance, *Macaca,* which has a number of species with fairly different locomotor habits, and for *Pan,* which has a pygmy variant that future research may show displays a difference in locomotion from the regular species), but at this time materials do not permit such a complete study.

---

2. Mixed = Dissecting room subjects.

TABLE 1

CANONICAL ANALYSIS OF SEVERAL ARRANGEMENT OF GENERA AND VARIABLES FOR TEST-
ING SUBSIDIARY HYPOTHESES. COMPARISON OF CANONICAL AXES ONE.   STUDY ONE--
DEFINITIVE ANALYSIS OF SEVEN SELECTED GROUPS AND EIGHT SELECTED DIMENSIONS.
STUDY TWO--EFFECT OF SEPARATING THE SUBGROUPS OF MAN ON THE EIGHT-DIMENSIONAL
STUDY.   STUDY THREE-EFFECT OF DOUBLING THE NUMBER OF VARIABLES.

Number	Groups	Values in Canonical Axes		
		Study 1 Above	Study 2 Above	Study 3 Above
29	*Macaca*	-8.4	+8.5[b]	-10.0
13	*Pongo*	-0.8	+0.9	- 0.1
98	*Pan*	+0.4	-0.4	- 0.2
98	*Gorilla*	+4.1	-4.0	+ 3.7
46	MAN	+0.7	--	- 1.2
16	Jericho	--	-1.2	--
14	Modern[a]	--	-0.7	--
7	Australian	--	-0.9	--
4	Vedda	--	+0.5	--
5	Andamanese	--	-0.3	--

[a]Specimens from an anatomical laboratory, England.

[b]Note that all figures in column two are reversed in sign as compared with
columns one and three. This commonly occurs in canonical analysis and, as
long as it applies to all items, signifies nothing. The relative arrangement of the items is not changed. See also table 2.

TABLE 2

CANONICAL ANALYSIS OF SEVERAL ARRANGEMENT OF GENERA AND VARIABLES FOR TEST-
ING SUBSIDIARY HYPOTHESES. COMPARISON OF SECOND CANONICAL AXES. STUDY ONE--
DEFINITIVE ANALYSIS OF SEVEN SELECTED GROUPS AND EIGHT SELECTED DIMENSIONS.
STUDY TWO--EFFECT OF SEPARATING THE SUBGROUPS OF MAN ON THE EIGHT-DIMENSIONAL
STUDY.   STUDY THREE-EFFECT OF DOUBLING THE NUMBER OF VARIABLES.

Number	Groups	Values in Canonical Axes		
		Study 1 Above	Study 2 Above	Study 3 Above
29	*Macaca*	-2.1	+1.9	-1.6
13	*Pongo*	+0.6	+2.0	-1.3
98	*Pan*	-1.1	-0.9	-0.6
98	*Gorilla*	-1.0	-1.1	-0.6
46	MAN	+8.7	--	+10.9
16	Jericho	--	+6.1	--
14	Modern[a]	--	+6.4	--
7	Australian	--	+6.4	--
4	Vedda	--	+5.7	--
5	Andamanese	--	+5.1	--

[a]Specimens from an anatomical laboratory, England.

TABLE 3

CANONICAL ANALYSIS OF SEVERAL ARRANGEMENT OF GENERA AND VARIABLES FOR TEST-
ING SUBSIDIARY HYPOTHESES.   COMPARISON OF THIRD CANONICAL AXES. STUDY ONE--
DEFINITIVE ANALYSIS OF SEVEN SELECTED GROUPS AND EIGHT SELECTED DIMENSIONS.
STUDY TWO--EFFECT OF SEPARATING THE SUBGROUPS OF MAN ON THE EIGHT-DIMENSIONAL
STUDY.   STUDY THREE--EFFECT OF DOUBLING THE NUMBER OF VARIABLES.

| Number | Groups | Values in Canonical Axes | | |
		Study 1 Above	Study 2 Above	Study 3 Above
29	*Macaca*	+0.6	+1.3	-0.2
13	*Pongo*	+3.9	+3.0	+6.8
98	*Pan*	-0.4	+0.1	-1.1
98	*Gorilla*	-0.2	-0.0	+0.1
46	MAN	+0.2	--	+0.0
16	Jericho	--	-1.0	--
14	Modern[a]	--	-0.1	--
7	Australian	--	-0.2	--
4	Vedda	--	+0.1	--
5	Andamanese	--	-0.4	--

[a]Specimens from an anatomical laboratory, England.

The results of the multivariate study of these data confirm that the different groups
of man are relatively very similar (even though they may be distinguished). This
first information is summarized by the canonical variates shown in tables 1, 2, and 3
(column 2) in which the human groups cluster together in all axes. This study also
confirms that the inclusion of man as a series of separate groups does not distort
the position of man as a whole when seen against the background of groups of nonhuman genera. This second information is provided in a study of tables 1, 2, and 3
(columns 1 and 2), where the relationships with other species are clearly similar
(although not identical) whether or not the human subgroups are pooled.

Another problem in these studies relates to the matter of how many dimensions it
is useful to examine. The existence of eight additional dimensions (to a total of 16)
for some of our groups is of value in testing certain subsidiary matters that are
pertinent to any study of this type but which frequently cannot be carried out. Although, for instance, eight dimensions are a reasonable number of variables to characterize each specimen (especially for a bone as small and uncomplicated as the
talus—see the scapula, Oxnard 1973a, and compare the skull, Howells 1973a), it is
nevertheless useful to analyze a larger number of measurements to see if inclusion of
extra data makes any major difference to the relative similarities of at least those
genera for which more measurements are available. This can help to provide an
additional degree of confidence in the results of the analysis of the smaller but more
wide-ranging data set.

In the case of the skull, as many as seventy dimensions may be used (Howells
1972). For a less complicated region like the scapula, although seventeen dimensions
are available and have been analyzed, it is clear that the major information lies

within the first nine (Oxnard 1973a). For the pelvis only nine variables have been examined, but it seems rather likely that examination of an additional suite will provide important new information—this bone is just more complicated than the scapula.

For the talus therefore, a subsidiary analysis of sixteen dimensions was carried out for the genera *Homo, Gorilla, Pan, Pongo,* and *Macaca,* for which reasonable samples are available. This allows us to evaluate what may be lost from the main analysis because it comprises a smaller number of dimensions (eight) taken on each specimen. The results show succinctly that, although somewhat greater separations are indeed achieved by the larger sample of dimensions, the pattern of these is entirely similar to that obtained from the analysis of the larger number of genera with a smaller number of dimensions (compare tables 1, 2, and 3, column 3).

Finally, one most important effect, that of the artificial manipulation of variables, has been tested in this particular study. Basic data for multivariate studies frequently consist of indices and angles (e.g., Ashton, Healy, and Lipton 1957; Ashton, Healy, Oxnard, and Spence 1965; Bilsborough 1971; Day and Wood 1968, 1969; Oxnard 1973c; and Zuckerman, Ashton, Flinn, Oxnard, and Spence 1973); such derived dimensions are chosen partly because it is thought that they may contain biomechanical information about the shape (e.g., may provide data about some biomechanical feature such as a lever arm), partly because they are crude attempts to eliminate size differences from analyses making more comparable large and small species, and partly because they summarize information by reducing two dimensions to one. Although opinions can be provided as to why the use of indices may be superior to measurements, it is clear that arguments can also be adduced relating to the dangers of using indices instead of measurements. Thus, possessing both the original measurements and the concocted indices for this study, it is possible to investigate the effect of such data manipulation in the derived multivariate statistics. Investigations of the effects of sex and age are also possible in this same subsidiary analysis.

All of these effects are studied in the third series of multivariate analyses in the following manner. Analyses are performed on that set of data which includes those genera for which sex and age are known, together with those genera for which the latter information is absent but which nevertheless contain the same set of variables. This comprises an analysis of eight dimensions for the genera *Homo, Gorilla, Pan,* and *Macaca,* for which large enough sex and age subgroups are available, together with additional genera which all happen to have the same measured variables and which help to provide computational stability. The analyses that are available consist of (1) a study of the eight measurements when combined in the form of a series of indices consisting of pairs of anatomically related characters, (2) an investigation of the eight raw measurements from which the indices are derived, and (3) an analysis of the eight measurements after a transformation has been applied to render less obvious the effect of overall size upon the variances of the dimensions within the groups (larger dimensions having significantly larger variances).

Figure 47 shows that when the data are examined as a series of indices, overall size between genera can best be defined through the projection of the groups upon a single axis (the large arrow) that lies at an angle to each of the first three canonical axes. The size differences within the genera (shown by the smaller arrow within each circle in the figure) are also obvious but lie at a series of different angles within the first three canonical axes for each individual genus. Again, within each genus the positions of specimens of each sex and each age are delimited in a general way as shown by the quadrants (young males, young females, old males, old females) or halves (males, females) within each genus in figure 47. The arrangement of the sex and age effects are not parallel from group to group.

However, examining a series of raw measurements (from which the indices had previously been derived) demonstrates (fig. 48) that the between- and within-size differences and sex and age differences are still preserved. This analysis also shows that considerably more information resides in the raw measurements than in the indices, because the overall separation between the genera is considerably greater. But the most obvious difference between this and the previous analysis is that, in the examination of raw measurements, the effects of the between- and within-genus size differences have been rotated within the canonical space (a) so as to be maximized within the plot of the first two canonical axes and (b) so as to lie approximately parallel to one another. In addition the discriminations between the within-genus sex and age differences have also been rotated so that they now lie parallel to one another as one passes from genus to genus, rather than at a variety of different angles as in the previous study.

The effect of logarithmically transforming the data before multivariate analysis produces overall separations between the genera that are even greater than those

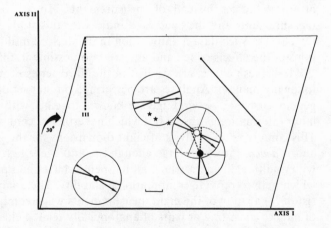

Figure 47. Preliminary canonical analysis of data taken on the talus in the form of indices and angles. Size seems to be expressed between the genera as shown by the large arrow that lies at an angle to each of the first three canonical axes. Within each genus, size is shown by the small arrows which point in the same general directions but show considerable differences from genus to genus. For those genera where sex is known the separation between males and females is indicated by the thick line within the generic circles. For those genera where age is known (young adults versus old adults) the circles are further subdivided by a thin line. The arrangement of the sex and age effects are also generally somewhat similar but do differ from genus to genus. Squares = macaques; solid circle = gorillas; open circle = chimpanzees; solid circle enclosing a star = man; solid stars = certain additional genera for which sex and age information is not known. The curved arrow and the 30° sign are an attempt to indicate that these differences really lie within three canonical axes.

obtained from the analysis of the raw measurements (fig. 49). But again the most striking change produced by this modification is the further rotation of the between- and within-genus size effects, and the within-genus sex and age effects, so that these now all relate to the separation achieved within a single canonical axis (the first), although it is clear that size is not the only information included in this axis. This represents a further simplification of the overall result, and it is for this reason that the final analyses (although carried out and available for all three modes) are here presented as derived from measures rather than indices, and from logarithmic transformations rather than raw measurements.

The Ultimate Study

As a result of these preliminary studies, the ultimate investigations upon which the biological conclusions are based comprise the univariate and multivariate analyses of eight[3] logarithmically transformed measurements taken upon each specimen of each genus.

This analysis thus includes the following groups: *Homo* (mixed, Jericho), *Gorilla*, *Pan, Pongo, Hylobates, Colobus, Macaca, Papio, Ateles*, together with the fossil groups *Proconsul* and "*Limnopithecus*,"[4] and the single fossil specimens for *Homo neandertalensis* (Kiik-Koba), *Australopithecus africanus* (= "*Homo habilis*," Olduvai), and *Australopithecus robustus* (= *Paranthropus*, Kromdraai).

Examination, in a univariate manner, again using basic statistics such as the mean and variance, show that, as Day and Wood found, *Gorilla* and *Pan* lie close to one another, as do the various groups of man to one another. But the inclusion of a number of other species shows that there is much overlap among the species, although the nature of this overlap varies markedly from dimension to dimension. When the various fossils are interpolated into these results, it is immediately apparent that they are not, in the case of the nonhuman fossils, similar to the African apes; neither are they, in the case of the single Neandertaler, similar to man; and the australopithecine fossils are not like either man or African apes; nor do they fall intermediately between them in many respects.

Study of means and standard deviations of the raw measurements for the ranges of genera examined here demonstrates that most dimensions tend to separate the

3. Only four measurements could be taken upon the specimen of *Australopithecus robustus* (= *Paranthropus*), and accordingly a series of estimates for its final measures are evaluated. When these measurements are assigned the values for *Australopithecus africanus* (= "*Homo habilis*"), the results are by far the most concordant with the entire study. However, replacing the missing measurements by those derived in turn from *Proconsul, Limnopithecus,* or *Pongo* has effects that differ scarcely at all from these. Replacing the missing measurements with those of any other genus produces a position for the fossil that does not lie in the same part of the canonical data-space as any of the examined primates. Four missing measurements in a large analysis such as this are most unlikely to have any biasing effect upon the final result. Certainly carrying out the analysis without this fossil produces an identical answer.

4. Further study of additional fossil specimens that are neither human nor australopithecine raises real doubts as to the reality of these two fossil generic groupings. This is more fully discussed in Lisowski, Albrecht, and Oxnard 1975. These particular specimens almost certainly include representatives of a wider primate fauna than indicated here.

Figure 48. This figure differs from the last only in that it is obtained from an analysis in which measurements are utilized rather than the indices and angles built up from measurements. The difference is that the within- and between-size effects and the within-genus sex and age effects have been rotated so that they are generally parallel and expressed obliquely within only the first two canonical axes. Conventions are the same as in figure 47.

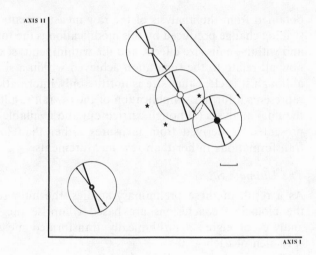

Figure 49. This figure differs from the last only in that it is obtained from an analysis in which the same set of measurements are logarithmically transformed. Here, between- and within-genus size and within-genus sex and age effects have all been further rotated so that they lie mainly in relationship to a single canonical axis, the first. Young females and old males are respectively the smallest and the largest specimens within each genus. Conventions as in figures 47 and 48.

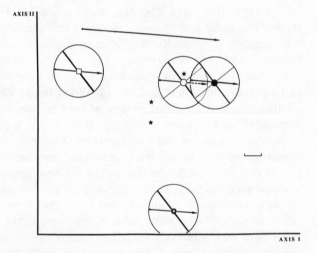

genera according to the overall size of the animals. Thus a rank order is usually preserved from the largest, *Gorilla*, through *Macaca*, the smallest, with various intermediate genera occupying generally appropriate intermediate positions. Examination of the variation within each genus for which we possess a large sample suggests that in addition, large specimens tend to be differentiated from small ones, males from females and young adults from old adults. Univariate tests demonstrate that many of the differences are statistically significant, although there is much overlap among the subgroups of specimens of each genus. Viewing the data for both indices and raw measurements, however, also reveals our inability to comprehend the totality of the information presented when only standard univariate analyses are performed, and, if for no other reason, a multivariate statistic would be of value.

Disposition of Extant Genera

The main multivariate analysis thus presents us with a more complete picture of the morphometric relationships existing among tali of different higher primates than was provided by Day and Wood (1968, 1969). Figures 50 and 51 give the principal results of the multivariate analysis of the new eight dimensions of the talus. The bivariate plot (fig. 50) of canonical axes one and two which together provide 70 percent of the information shows that man is clearly and uniquely separated by some seven standard deviation units from the various nonhuman genera. This separation resides almost entirely within canonical axis two. The nonhuman genera are themselves well separated by canonical axis one over a spectrum of some fourteen standard deviation units. Thus the African apes lie at one end of the separation of axis one. *Papio* and *Ateles* lie in the intermediate region of this axis, and *Colobus, Hylobates,* and *Macaca* lie at the opposite extremes. Some separation among the nonhuman genera is achieved by canonical axis two, but only to a small degree (at a maximum of five standard deviation units).

In addition to the separations achieved by the first two canonical axes, considerable information (13 percent) is also contained in canonical axis three. Figure 51 shows that this axis produces no further separation of man from other genera, but increases the separation among the nonhuman primates so that these can no longer be thought of as lying in a single linear system. Thus although most of the nonhuman genera are scarcely separated by this axis, two of the intermediate genera are markedly distinguished. *Pongo* is an outlier in the positive direction in canonical axis three and *Ateles* in the negative.

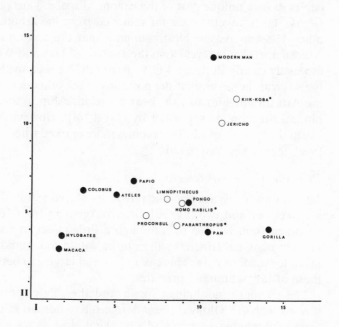

Figure 50. Bivariate plot of canonical axes one and two for the ultimate analysis of eight variables taken upon the talus in a range of primates. Solid circles = extent forms; open circles = fossil forms.

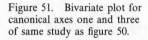

Figure 51. Bivariate plot for
canonical axes one and three
of same study as figure 50.

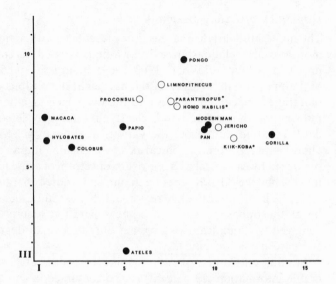

The analysis makes it clear that the various arboreal genera form a multidimensional structure incorporating almost all of the groups examined and having as its most extreme members (i.e., lying at the apices of the overall separation of the arboreal forms) the genera *Pongo, Hylobates, Ateles,* and *Macaca* (fig. 52 and 53). The various other genera (not shown in the figure) fall within this envelope. Two groups of terrestrial animals are outliers to this main arboreal group and each occupies its own unique part of the canonical space. Thus (a) *Homo,* and (b) *Pan* and *Gorilla,* taken together, are far removed from the arboreal species and from each other. It seems reasonable to suppose that the separation between man and the African apes that is revealed in the analysis of Day and Wood (1968, 1969) and seen especially clearly in figure 4 of Oxnard (1972c) is equivalent to that noted here. (It is also clear, however, that the positions of the other genera, when interpolated into the African ape-human axis, bear no relationship to any presumed quadrupedal-bipedal function [as suggested by Day 1974]). However, the real resemblance between the two sets of studies becomes more marked when the positions of the various fossil genera are considered.

Disposition of Fossil Genera

The various fossils are also unequivocally placed within the bivariate plot of canonical axes one and two (fig. 50). Bronze Age man from Jericho and the Neandertal specimen from Kiik-Koba, although a little closer to nonhuman primates than is modern man, nevertheless still both lie within four standard deviation units of the mean for modern man. There is a clear and large gap between these specimens and those of the nonhuman primates.

The obviously nonhuman fossils and the single specimens of *Australopithecus* ("*Homo habilis,*" Olduvai; *Paranthropus,* Kromdraai) fall squarely within the spectrum of nonhuman forms and somewhat close to its middle, so that the nearest

Figure 52. Generalized
distance analysis of the same
eight variables taken upon
the talus in a range of
primates as in the previous
two figures. The model is
constructed in three
dimensions from the distances
in the generalized distance
matrix. Only those genera
that are most peripheral are
shown. Thus man and the
African apes (represented
here by gorilla) are rather
outlying to the main envelope
of genera bounded by
orangutan (*Pongo*), macaques
(*Macaca*), spider monkey
(*Ateles*), and gibbon
(*Hylobates*). Within this
envelope are included all of
the other genera examined.
The general scale of this
diagram is 20 distance units
from side to side, 16 distance
units from top to bottom and
12 distance units in depth.

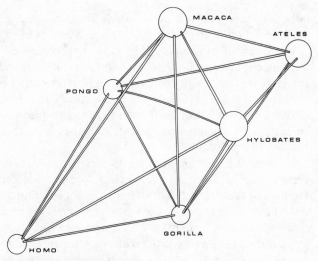

Slight bending of the
connections indicates where
higher dimensional distances
were compressed or forced
into three dimensions in the
construction of the model.

Figure 53. This is another
view of the model of figure
52, which has been turned
through an angle so that the
triangle of separations
between man, the African
apes and the orangutan now
lies at right angles to the
plane of the paper. It shows
particularly clearly the
volume bounded by the
orangutan, the macaque, the
gibbon, and the spider
monkey within which lie the
remaining genera that have
not been plotted.

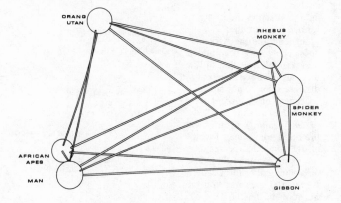

genera appear to be *Pongo* and *Pan*. Again, in the first two axes the separation of
the fossils from man is unequivocal (fig. 50).

The third axis (figure 51) does not significantly add to the separation of Bronze
Age man (Jericho) or Neandertal man (Kiik-Koba) from modern man (mixed).
However, canonical axis three does tend to separate all the fossils except Kiik-Koba
somewhat from the main body of extant nonhuman primates in the direction of the
outlying (as viewed in axis three) genus *Pongo*. This effect removes these genera
from proximity to *Pan*.

The disposition of Bronze Age man (Jericho) and the Neandertal (Kiik-Koba)
specimen, when viewed in relation to the total generalized distance matrix, is entirely

in accordance with the idea that these are human remains. They lie within five standard deviation units of the mean for modern man (fig. 54a and 55a).

The locations of the pooled groups of specimens of nonhuman fossils including the single individuals of *"Homo habilis"* (Olduvai) and *Paranthropus* (Kromdraai) are in accordance with the previous reinterpretation of Day and Wood's original result by Oxnard (1974c), and they are clearly foreshadowed by the more restricted study of Lisowski and Darlington (1965) and Lisowski (1967). All of these fossils lie further from man and the African apes than these two groups do from one another (fig. 54a and 55a).

But a most important addition to these previous results is that this degree of separation and its direction are such as to place the genera within the envelope of arboreal species that has just been described; more specifically, they are placed reasonably close to one of the arboreal species, the orangutan. The similarity of this result with that of Oxnard (1972c) can be seen by comparing a figure from that paper (reproduced here as fig. 54b) with that drawn from this study (fig. 54a). The distances and directions of these fossils from man and the African apes, although not absolutely identical, are similar enough to be most striking.

Figure 54.   The first frame (a) is a more detailed view of a small part of the model obtained from the same analysis as the previous four figures. It shows the triangle formed by man, the orangutan, and the African apes. There have been added the various fossils that fall in this part of the generalized distance space. Thus single specimens of *Australopithecus* (*Paranthropus* and so-called *"Homo habilis"*) together with the other non-human fossils all fall somewhat nearer to the orangutan than they do any other extant form. The Neandertal specimen and the mean of the Jericho specimens (not labeled) fall rather close to man. This can be compared to the model shown in the lower frame (b) that is constructed by Oxnard (1972c) from the generalized distances in the study by Day. Although in this case the orangutan is not present, and although the Neandertalers used are different specimens, the similarity with the current study is most marked.

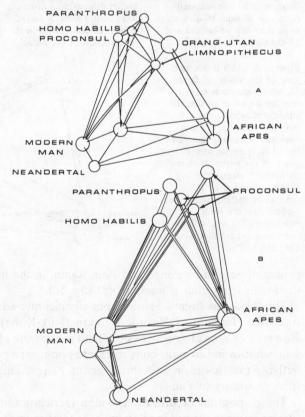

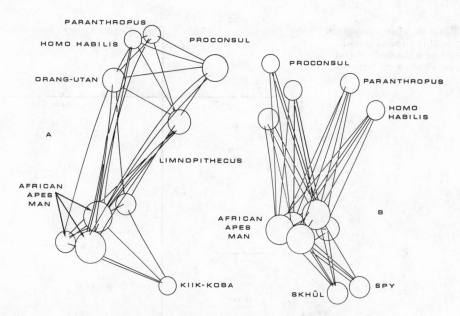

Figure 55.   The first frame (a) of this figure shows the model of the first frame of the previous figure turned through 90° and confirms the reality of the similarities between the Neandertalers, Jericho specimens, and modern man, and between the australopithecine, proconsuline, and "limno-pithecine" specimens and the orangutan. But the main point of this figure is the marked similarity with the second frame (b), which shows the rotated view of the generalized distance model result-ing from Oxnard's (1972c) reinterpretation of the study by Day. Again, although there are some differences between the two figures, it is certain that, despite different measurements and different specimens being used in the two investigations, the resulting arrangement is generally extremely similar.

This can be seen even more clearly when the data are turned through an angle so that man-African ape separation is viewed along its axis at right angles to the plane of the paper. The Neandertal specimens measured by Day and Wood (1968) fall, in the reinterpretation by Oxnard (1972c), downwards and to the right in the re-produced diagram (fig. 55b). This should be compared with an equivalent position for the specimen from Kiik-Koba in the current analysis (fig. 55a). In a similar manner, the specimens of *Proconsul* and those from Olduvai and Kromdraai ex-amined in the previous study fall towards the upper right and further away in the reproduced figure (fig. 55b). The rotated model of the present study (fig. 55a) places *Proconsul* together with the Olduvai and Kromdraai specimens in the same general part of the picture. The similarity is not complete; one would scarcely expect it to be so, because there are many more groups in the present analysis and different measurements are taken on each specimen. But the overall picture is close enough to confirm that we are truly dealing with an equivalent data-space in the two studies.

We may well ask if all this information is contrary to that published by Day in his original study. Careful examination of Day and Wood's (1968) original results (fig. 56) and comparison with Oxnard's (1972c) reinterpretation (fig. 57) suggests that it

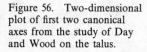

Figure 56. Two-dimensional plot of first two canonical axes from the study of Day and Wood on the talus.

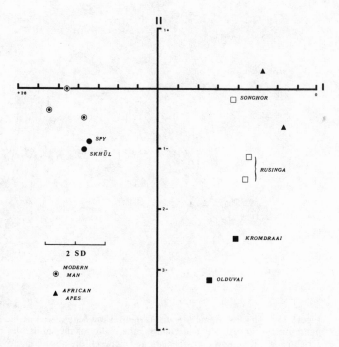

is not. Although Day and Wood's study is based upon data that differ somewhat from those used in the present analysis, the general picture is the same. Their speculations, of course, differ, being based only upon the first canonical axis.

High-Dimensional Information

Yet further separations of note are achieved by even higher canonical axes and these are best displayed through the use of the technique of Andrews (1972, 1973). First the method reiterates the unique position of man that is also clearly seen in the bivariate plot of the first two canonical axes. But it also confirms that the apparent linear arrangement of the remaining nonhuman extant primates is, when higher dimensional information is considered, more complicated. Thus the plot of functions (fig. 58) shows that the spectrum of separation is represented at the periphery by five distinct modes: (1) the Old World monkeys, (2) the African apes, (3) *Hylobates*, (4) *Ateles*, and (5) *Pongo*. This is foreshadowed by the additional information in canonical axis three just discussed.

Examination in the same way of the various fossil groups confirms that Bronze Age (Jericho) and Neandertal man (Kiik-Koba) are markedly similar to modern man (mixed) and quite separate from any nonhuman primate (fig. 59).

Examination of the functions for the remaining fossils including the individual single specimens of *Australopithecus* confirms and strengthens the suggestion of canonical axis three that these genera are most similar to *Pongo* (fig. 59). However,

Figure 57. Three-dimensional model constructed by Oxnard on the basis of the analysis of the previous figure. The first frame (labeled) shows a view of the model that entirely corresponds with the two-dimensional figure of Day and Wood. Succeeding frames show rotation of the model on a horizontal axis until at the bottom we come to the view suggested by Oxnard that displays the ways in which the various fossils differ from the extant groups. This is achieved by utilizing the separation between man and the African apes as the axis of rotation.

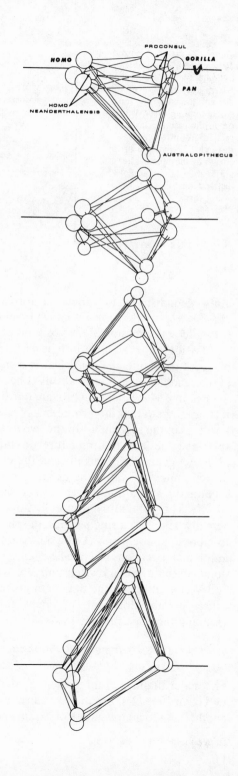

Figure 58. High-dimensional display of the analysis of the data on the talus. Extant genera only are shown either through the curve for the corresponding mean value or, in the case of closely related groups of genera, through the envelope of the curves of the mean. The graph shows the complete disparateness of each extant form.

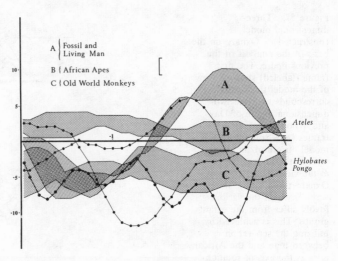

it also demonstrates that they fall in two pairs: *Proconsul* is remarkably similar to "*Limnopithecus*";[5] "*Homo habilis*" is much like *Paranthropus* (fig. 60).

Finally, examination of the positions of individual specimens of the various extant and fossil groups is shown in figure 61. Here it is clear that each extant genus (displayed only within the canonical two-space but also confirmed within the totality of the canonical axes) forms a truly coherent group. The dispersions of the specimens for *Pan* and *Gorilla* are slightly greater than for the other genera, but this is clearly related to the rather larger number of specimens in those two groups. However, the positions of the specimens for the two fossil groups for which we have more than one specimen does not conform entirely to this picture. The seven specimens of "*Limnopithecus*" generally lie together in the canonical two-space, being spread only to a degree similar to that of the extant genera. But the five specimens of *Proconsul* are separated by an amount greater than that found in the ninety-eight specimens for *Gorilla*. Thus while it is not totally impossible that the five specimens of *Proconsul* actually belong to a single genus, the chance also exists that these specimens belong to more than one genus (Lisowski, Albrecht, and Oxnard 1974b, 1975). Indeed two individual specimens of *Proconsul* fall somewhat towards *Macaca* and *Papio,* a result quite in accord with a revised study by Wood (1973) of some of the talus data of Day and Wood, although the data is interpreted here considerably more cautiously because enough additional specimens are available to question the reality of the grouping (and see further, Lisowski, Albrecht, and Oxnard 1975).

Morphological Meaning of Relationships
between Fossil and Extant Genera

The major findings of the present analysis, therefore, over and above that reported by Oxnard (1972c), are that we can now place these fossils in relation both to one another and to a larger suite of extant species. The general morphological similarity

5. See footnote 1 of this chapter.

Figure 59. High-dimensional displays for particular genera in the canonical analysis of the eight dimensions taken on the talus. The first frame (a) shows the similarity of the Neandertal specimen and the envelope for man, together with the extreme difference of the australopithecine specimens from man. The second (b) and third (c) frames show the differences of the australopithecine fossils from the envelope for Old World monkeys and African apes. The third frame shows the marked similarity between the plots for the australopithecines and the orangutan (*Pongo*). This confirms the information from the previous displays.

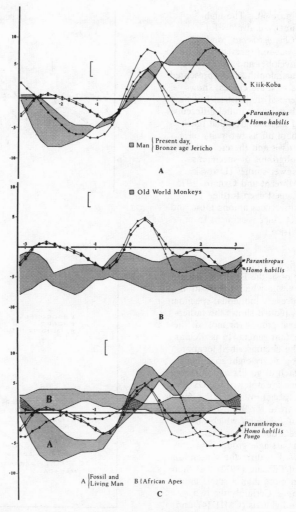

(though not by any means of identity) of those nonhuman fossils that are considered is with the arboreal ape *Pongo*. This is true whether we look at the first three canonical variates alone (fig. 50 and 51), the three-dimensional models (fig. 54 and 55), or the full high-dimensional analysis (fig. 58, 59, 60). What may be the meaning of this relationship?

First, it is presumably possible that the finding is accidental. If that proves to be the case, then we can read no biological meaning into the resemblance to *Pongo*. The only speculation that would remain is that of Oxnard (1972c) that these fossils, most especially "*Homo habilis*" and *Paranthropus* are merely uniquely different from both the African apes and man.

The nature of the relationships among the extant arboreal species suggests rather strongly that the shapes of the different tali reflect the various mechanical situations

Figure 60. The high-dimensional plots for *"Limnopithecus"* and *Proconsul* (presented as an envelope) and for the two australopithecines (presented as separate curves) show general similarities. But they also show parallel differences of lesser degree that mark them off as two pairs of curves and therefore two subgroups of specimens. Newer studies (Lisowski, Albrecht and Oxnard, 1975 suggest even further differences among these and yet other specimens from Africa.)

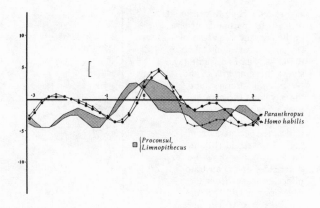

Figure 61. The bivariate plot of the first two canonical axes in which the position of every individual specimen is plotted shows the rather neat groups for most of the extant genera. In particular the circumscribed locations for ninety-eight specimens each of gorilla and chimpanzee are noteworthy. Of especial interest is the curious spread for the five specimens of *Proconsul* which throws real doubt upon their reality as a single group. This may well confirm the contention of Pilbeam (1972) that there is more than a single entity here. Certainly the two individuals (CMH 147 and 234'50) found by Wood (1973) in a reevaluation of his own data, do fall somewhere near to *Papio* (but see Lisowski, Albrecht, and Oxnard 1975).

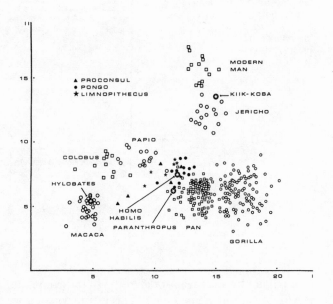

of that bone within the locomotor contexts of these vastly different animals. Thus, of the species examined here, it is quite clear that the slow acrobatic climbing of the orangutan, the fast brachiation of the gibbon, the acrobatic arm- and tail-swinging of the spider monkey, and the regular running, jumping, and climbing of the Old World monkeys (whether in the trees or on the ground) represent different polarities of locomotion. These locomotor differences seem to be reflected in the differences among the species in the form of the talus. If this is so, then the positions of the nonhuman fossils suggest that they are unlikely to have been creatures possessing tali modified for regular arboreal or terrestrial running and leaping like the various Old World monkeys examined here (save perhaps for some specimens currently assigned to *Proconsul*). Nor are their tali modified in a way that indicates use of the hindlimb associated with ricochetal brachiation as in the gibbon, or modified brachiation in the spider monkey. Clearly also these bones are not adapted in a manner similar to those of man (a terrestrial biped) or the African great apes (terrestrial knuckle-walkers).

The possibility must remain that, in the shape of their talus, the nonhuman fossils may be reflecting functions of the foot that may relate to acrobatic arboreal climbing such as is reminiscent of the extant species *Pongo*.

This statement must, of course, be qualified. It is highly unlikely that these fossil creatures actually moved in a manner identical to the slow and careful climbing of the very heavily bodied adult male orangutans. The marked difference of body weight must be taken into account. The morphological similarities may, however, mean that the animals may have been capable of moving as intermediately sized decidedly arboreal creatures but without the extreme quadrumanal acrobatics of the massive orangutan. The foot of such an animal as evidenced by the talus may well have participated in climbing activities within the trees. Pending further evidence we are left with the vision of intermediately sized animals, at home in the trees, capable of climbing, performing degrees of acrobatics and perhaps of arm suspension sufficient to relate to frequently orthograde trunk positions but without the extreme hip mobility of the orangutan, and certainly without the ricochetal brachiation of the gibbon. Presumably such an animal may have been moving in a way similar to what would be expected of the arboreal ancestors both of man and the African great apes.

A speculation like this may be thought entirely reasonable for the earlier forms: the specimens of *Proconsul*. But for *Paranthropus* and especially for "*Homo habilis*," animals which are generally presumed to have been bipedal, the current findings may be thought to be curious at the very least. However, if we assume that skeletal shape reflects not only what an animal does during its own lifetime, but also to some degree what it has done in its own recent evolutionary past, then we can see a possible reason why these latter forms fall where they do. Such a reason—retention of a degree of climbing adaptation—excludes any morphological evidence of a knuckle-walking or richochetal brachiating stage in their relatively recent ancestry. An alternative explanation, is, of course, that *Paranthropus* and "*Homo habilis*" were not bipedal in the manner of man, their terrestrial locomotion perhaps resembling more the bipedality found in a number of extant primates, especially, for instance, the chimpanzee. Presumably, the well-known pelvic evidence controverts this idea, although it seems that this evidence is no longer as strong as it once was though to be (Zuckerman, Ashton, Flinn, Oxnard, and Spence 1973; see also chapter 3 of this volume).

A Terminal Toe Phalanx

The foregoing studies are all from our own laboratories. An important test of the major inferences that they have provided so far (i.e., the uniqueness of the australo-pithecines and especially of their lack of resemblance to man and the African apes) must also depend upon other workers making similar findings. However, there are already a number of multivariate studies of different parts of the body that may be applied as appropriate tests.

For instance, one study in which the multivariate approach has been utilized is of the hominid terminal toe phalanx (Olduvai Hominid 10) recovered from Olduvai in Tanzania (fig. 62). Observational evidence has been put forward to support the view that this bone belongs to an "upright hominid with a plantigrade propulsive or striding gait." The subsequent multivariate study carried out by Day (1967) utilizes as raw data a series of nine characters that he believed to be functionally significant. They comprise: maximum length, length-breadth index, minimum mid-shaft thickness, index of robusticity, the breadth of the head, the convergence angle (a measure of the relationship between head-breadth and base-breadth), and the angles of axial deviation and of axial torsion of the shaft.

Although these measures are chosen with the idea that they are functionally significant for locomotion—and there can be little doubt that this is so—it is also clear that they must also reflect many other aspects of the shape of the toe bone: for instance, features associated with growth, with genetic heritage, with nutrition and disease, and so on. And of course this is the case for dimensions of the pelvis, talus, and scapula previously discussed. We should also continue to bear in mind that for most of these studies no experimental work is available to confirm the functional significance of the parameters. It is only for the scapula that a precise study has confirmed the mechanical relevance of specific elements of the shape that were measured in those studies. Thus the measurements of the toe bones, as with those of the other regions, are functionally significant for locomotion only to some undetermined extent; they are also significant in many other ways.

The data were obtained from a total of 122 toe bones from men, gorillas, and chimpanzees, and the measurements were compounded by means of generalized distance and canonical statistics. Since there are five groups (three of man and two of

Figure 62. Outline drawing
of the right terminal phalanx
of chimpanzee, fossil,
and man.

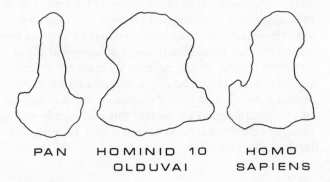

PAN   HOMINID 10   HOMO
              OLDUVAI   SAPIENS

African apes) in this analysis, there will therefore be only four canonical axes (one less than the number of groups) separating these groups however many variables are examined. Being familiar with the materials as we are, it is clear that the differences within the three human groups and the differences within the two African ape groups are of much smaller magnitude than the differences between apes and men. A child could not mistake an ape toe bone for a human toe bone, though it might well mix up different men with one another and the two apes with one another. Accordingly, we recognize that whatever analysis is attempted, we are making discriminations between only two major groups. Accordingly, they can only be separated by a single axis, whatever we do. Thus it is dangerous to assume that this single axis is only locomotor in meaning (i.e., separates bipeds from quadrupeds, as in Day 1967, 1974). This single axis also contains all of the other information that separates men from apes. This is not only the locomotor difference, but also the taxonomic one: it presumably also includes the nutritional difference, the allometric difference, the sexual dimorphic difference, and, as long-shots, the differences due to domestication and civilization, and perhaps even yet others that we have not identified and can-not even begin to guess at.

And if we attempt to interpolate some other bone—whether of man, ape, monkey, or elephant—into this axis, though it is clear that it will fall somewhere, it is not clear what the meaning of its position will be if it truly does not belong with one or the other of the original groups analyzed. Locomotor inferences from a projected position in this first axis are unlikely to be correct.

The placing of the Olduvai Hominid 10 into this analysis provides us with just such a problem. For this fossil is projected into the first canonical axis, as is shown by Day, in a position close to man (fig. 63). Construction of the true relationships from the generalized distances can be seen in a three-dimensional model (because we are comparing essentially three sets of differences—the first between men and apes, the second among men and among apes, and the third between the fossil and the rest). This model can be viewed in a way that confirms the arrangement of the extant groups and the projected position of the fossil into the plane of axes one and two close to man (fig. 64). But rotation of the model through almost ninety degrees using the ape-man separation as the axis of rotation clearly shows the unique nature of the fossil (fig. 65). Another rotation further confirms this (fig. 66). This has been confirmed independently by Benfer (personal communication), who has shown, using multidimensional scaling, that the Hominid 10 phalanx is located as far distant from the samples of *Homo sapiens* as these latter are from the African apes.

Having established that the fossil is quite different from man or the African apes, we must note that there remains, therefore, the possibility that Day (1967, 1974) is correct in the functional interpretation that he has drawn. Does the projected posi-tion of the fossil say something about the bipedal-quadrupedal relationship? First, we have just shown that the arrangement in axis one contains all of the differences between man and the African apes and not just those that are locomotor. But, further than this, we have a sound example, that provides evidence bearing upon this matter.

Figure 63. Canonical
analysis of dimensions of the
toe bone after Day. The
values on the axes are those
of Day (1967).

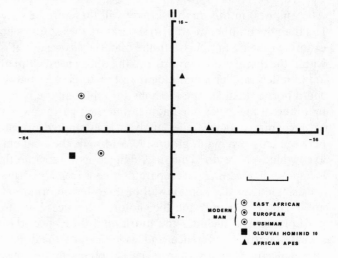

Figure 64. The three-
dimensional model con-
structed on the basis of the
generalized distances in
the analysis resulting in figure
63. Note the similarity in
position of everything in
this model as in the previous
figure. The general scale of
the model is 8 distance units.

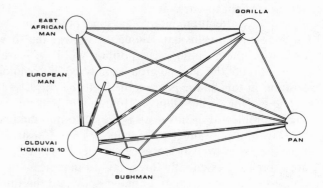

Figure 65. The same model
as figure 64 rotated through
90°. The fossil is actually
located as far distant from
man as man is from the
African apes.

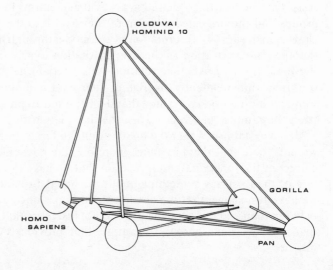

Figure 66. Another view of the model that even suggests that man could be thought to be somewhat intermediate between the African apes and the fossil.

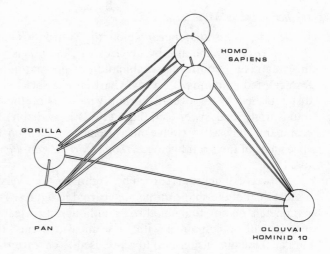

Thus, in the canonical analysis of data on the primate scapula, canonical axes one and two appear to separate the nonhuman primates in ways that make considerable sense from a locomotor point of view. However, Oxnard (1969a) showed that a prediction based upon this finding appeared to fail when the position of man is considered.

> It would suggest for instance, that man is capable of raising his arms and hanging from them to a degree reminiscent of the woolly monkey (they lie in similar positions in terms of axis one) and further, that he is more arboreal than any monkey or ape (man is more negative in axis two than any other form). However the general hypothesis is falsified only if the position shown for man is his true position. This is not the case. There are only two axes separating the monkeys and apes from one another; examination of the third axis as revealed in the bivariate plots of axes one and three shows that man is not co-planar with other forms; he actually lies considerably more negatively in axis three and is thus separated uniquely from the nonhuman primates.

An exactly similar finding applies in the current situation thus negating the argument of Day (1974). It is clear, therefore, that any functional interpretation of the Olduvai Hominid 10 phalanx from these data is problematical. All that we may say at this time is that the form of this bone in Olduvai Hominid 10 is inconsistent with the notion that it possesses adaptations related either to human or African ape feet as we know them today. It is still possible, of course, that future studies will show that it was from some kind of biped. Other primate genera are not available for comparison in this study, but, although further investigation may show similarities of adaptation with some other living form, as proved the case for the talus, the general picture presented by this fragment at this time is only that its uniqueness from man and African apes is not in doubt.

Higher Primate Metacarpals

In the case of this anatomical region, the Swartkrans metacarpal described in detail by Napier (1959) and also by Le Gros Clark (1967) is available. Napier has seen in this fragment a confusing combination of pongid and hominid features, with the former predominating. Le Gros Clark also associated the fragment with a hand distinguished by its overall "primitiveness." For the Swartkrans metacarpal the multivariate analysis of Rightmire (1972) is available. Rightmire performed a discriminant analysis of twelve measurements taken on the first metacarpal of 105 specimens of the extant genera *Homo, Gorilla, Pan, Pongo, Hylobates* as well as the fossil.

The twelve dimensions chosen by Rightmire are a very detailed attempt to capture a great deal of the shape of the metacarpal in figures. There is no attempt to choose parameters on any functional basis, although it is clear that much functional information will be contained with these measurements. It is of interest how closely similar these dimensions are to those used to characterize the shape of the toe bone in the study just described. Thus these measurements comprise: maximum length, mid-shaft subtense (a measure of longitudinal curvature of the shaft), mid-shaft angle (computed from the subtense to show curvature in degrees), proximal articular breadth, proximal medio-lateral curvature as obtained from subtense, transverse mid-shaft breadth, mid-shaft height, distal articular breadth, maximum head-breadth and distal articular height.

Rightmire's analysis provides a two-dimensional display of the first two discriminant functions which superficially appears to suggest human associations for the fossils (fig. 67). However, Rightmire is mindful of the possibility that higher axes may contain additional information, and his final conclusion denies any human associations in the morphology of the fossil; he suggests that "the fossil metacarpal may be functionally similar to that of the chimpanzee."

Figure 67.   The two-dimensional plot of the first two canonical axes of the study on the Swartkrans metacarpal by Rightmire. In this diagram the fossil looks rather similar to man but Rightmire is aware that other information shows that it is different.

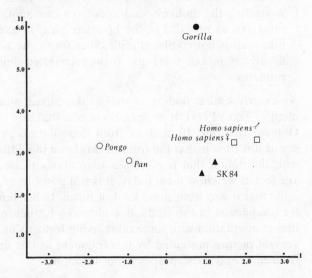

In this study sufficient information has been presented (Rightmire 1972 and personal communication) to allow a three-dimensional model to be built. Shown with minimum distortion, one view of this model is displayed in figure 68, and it confirms entirely Rightmire's two-dimensional plot of the canonical axes (compare with figure 67). Rotation of the model around an approximately horizontal axis shows, however, the extra information noted by Rightmire, and it indicates that he is technically correct: of the extant genera in the study, the fossil is marginally closer to the chimpanzee than to any other form (fig. 69). But the principal finding, hinted at by Rightmire, is that these fossils are, in truth, much further away morphologically from any of the extant specimens than these are from each other. Clearly the better conclusion is that the fragments are uniquely different from all of the extant types examined.

Rightmire has attempted to investigate further the "meaning" of his discriminant axes by looking at the differential contribution of his original variables to them. Thus, for the extant forms, axis one (60 percent of the information) separates the specimens largely through breadth measurements of the distal extremity of the bone and by overall length. Robusticity may be being registered here, as also may a lack of curvature. The second axis (29 percent of the information) appears to draw together measures of the base of the proximal articular surface and of high curvature, but the contributions are not easily understood in terms of the function of the bone. A third discriminant axis (but not, however, the third dimension in our model) contains 8 percent of the information and seems to be reflecting size.

The marked distance of the interpolated fossil from all of the extant forms is not related to any of these particular discriminant axes. Rather it is added due to the fact that the process of interpolation of a specimen that does not "belong" near to any of the original groups greatly increases its own generalized distance from the others. The fossil is unique.

Figure 68. The model that is constructed from the generalized distances that can be obtained from the previous study. Careful examination shows that each genus occupies approximately the same position in the two displays. The general scale of the model is 8 distance units.

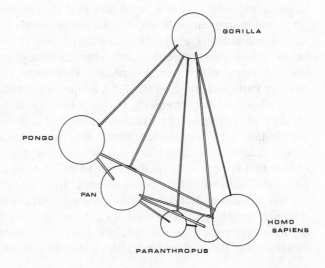

Figure 69. The model of figure 68 turned through almost 90°. This demonstrates that Rightmire is correct when he says that these fossils, though resembling man in the first two axes, are really closer to the chimpanzee. But the reality is even more distinct. The fossils, though truly closer to the chimpanzee than to anything else, are in fact so far even from that genus that they must be reckoned to be uniquely different from all living hominoids. The general scale of the difference between the fossils and the extant forms is 8–10 distance units, that among the various hominoids is not much greater than 4 units.

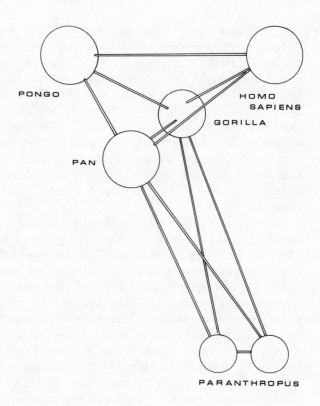

The Distal Humerus in Higher Primates

Several portions of fossil distal humerus are known from Africa; McHenry (1972, 1973) has supplied the appropriate multivariate morphometric studies. This particular anatomical region is important in the reconstruction of the locomotor, postural, and manipulative behaviors of early hominids because at some time or other its function must have undergone a change from the weight-bearing of nonhuman behavior to the non–weight-bearing characteristic of modern man. It is well known that the shape of this part is especially difficult to assess (Straus 1948, and more recently Patterson and Howells 1967, having noted the remarkable similarity of the various hominid distal humeri). This is, however, one of the two situations in which multivariate analysis is of particular value—that is, in making discriminations among objects that are quantitatively rather similar (the other situation in which multivariate analysis is of particular use is the converse; for assessing relationships among objects that appear to be qualitatively so different as to defy analysis, see the comparison of the innominate bones of man, apes, and monkeys).

McHenry has undertaken, therefore, a multivariate analysis on a total of eighteen measurements taken from a series of 221 specimens of the distal humerus representing man, chimpanzee, gorilla, and orangutan. The measurements include a series of dimensions of the trochlea (its width and height and the diameter of its lateral ridge), the capitulum (its width, height, and the width of its articular surface), the olecranon fossa (its width and depth and the width of its medial and lateral walls), the shaft (its antero-posterior diameter, its circumference, and its total length), and finally the epicondyles (their individual widths and the bi-epicondylar width).

These measurements are not chosen with any specific biomechanical ideas in mind (the anatomical region is rather small and the biomechanics may be too little understood to allow this; McHenry 1972); they comprise, however, a suite of measurements that surely characterize a great deal of the shape of this small anatomical region and undoubtedly do contain much functional information.

McHenry (1973) presents a three-dimensional view of the information in the first three discriminant axes, realizing that although the first two axes suggest interesting relationships between the Lake Rudolf fossils and certain extant groups, the third and to a smaller degree even higher dimensions also contribute information of value in discrimination. Accordingly, his final conclusion is, correctly, that this fossil distal humerus is unique among extant hominoids (fig. 70).

This conclusion is confirmed in the more extensive study described by McHenry (1972). In this latter investigation, a larger number of extant genera together with data from additional fossil specimens are included. The picture is now so complicated that neither a two-dimensional plot nor a three-dimensional model provides any very useful view. But if we assess the full analysis of McHenry, when all canonical axes are included, we can confirm clearly that the fossil from Kanapoi is very humanlike; this is not new. However, we can also confirm that the East Rudolf specimen is essentially unique and can suggest in addition that it has possible similarities to the orangutan. McHenry's own conclusion about the Kromdraai fossil is that it is rather more like man than any other extant form. However, we can see that it is indeed very near the tripartite boundary between man, the African apes and Asiatic apes, and it is also somewhat nearer to the Lake Rudolf specimen than to the mean of any extant group. A more conservative view might be that it is actually somewhat different from all three extant forms.

Figure 70. The two-dimensional plot that arises from the studies of McHenry on the distal humerus. In this diagram Kanapoi looks closest to the orangutan and Kromdraai to man; East Rudolf is different from any extant form. McHenry is aware, however, that higher axes produce other resemblances.

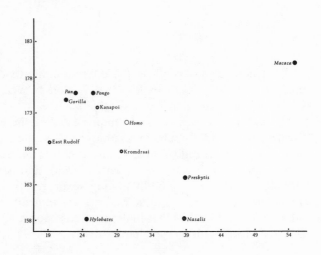

We have used the multidimensional technique of Andrews (1972, 1973) to demonstrate (figs. 71 through 76) these relationships as they apply to the East Rudolf and Kromdraai specimens. The multidimensional display demonstrates (a) the similarities that exist among all of the great apes through the conjoint envelope representing them, (b) the difference of the group of great apes from the conjoint envelope for man and the specimen from Kanapoi, and (c) the difference of both of these groups from the single plot representing the mean position of the lesser apes (fig. 71). That the Kromdraai and Lake Rudolf specimens differ markedly from each of these extant groups as well as from each other is shown by figures 72 and 73. In order to be sure that similarities with at least some of the Old World monkeys can also be excluded, comparisons with proboscis monkeys and langurs are also provided (fig. 74). If these latter genera have any similarities at all among the hominoids they are with the lesser apes (fig. 75), creatures with which they share some features that relate to acrobatic activity in the trees.

Figure 71. The high-dimensional display of the arrangement of extant genera in McHenry's study of the distal humerus. The complete difference between man, gibbons and great apes is clearly evident.

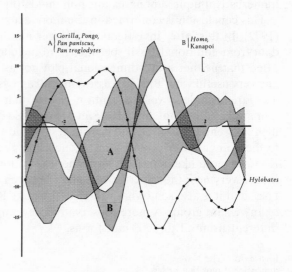

This technique also confirms the extreme similarity of the Kanapoi specimen and modern man (fig. 76), a finding that is already well known, of course, from the studies of Patterson and Howells (1967).

The different morphological uniquenesses of the Kromdraai and East Rudolf fossils may also imply functional uniquenesses; these early hominids are equipped with distal humeri unlike those of any extant form. But although Kromdraai is the more human, its direction of departure from man is towards the orangutan. And although East Rudolf is totally unique, its nearest living analog is again the orangutan. The significance of these findings are difficult to assess but, when viewed in relation to some of the other findings in this book, they are not so totally out of line as to be incompatible (see Chapter 6).

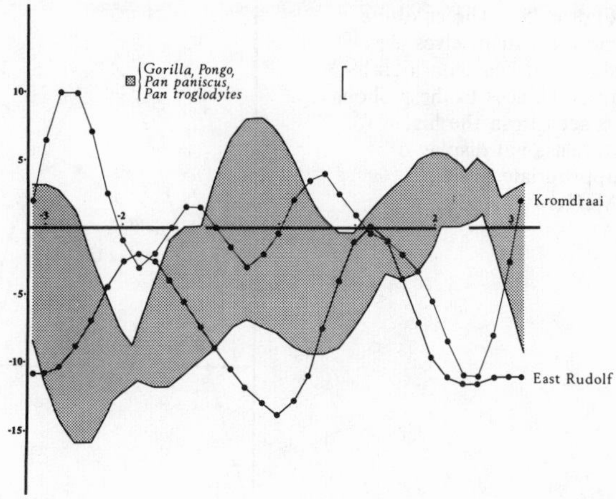

Figure 72. The two fossils examined by McHenry are clearly different from all of the great apes as shown by the high-dimensional display.

Figure 73. This high-dimensional display of McHenry's results shows that the two fossils are different from the envelope representing man.

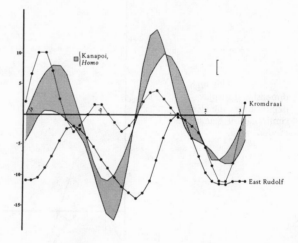

Figure 74. The high-dimensional display of results from McHenry's study on the distal humerus indicates that the fossils are different from the envelope of colobine monkeys.

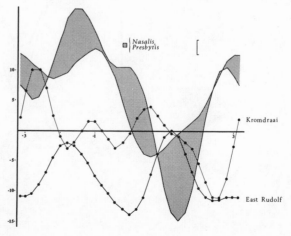

Figure 75.   The colobine
monkeys themselves are
shown to bear considerable
resemblances to the gibbons
as seen from the high-
dimensional display of
appropriate parts of
McHenry's study.

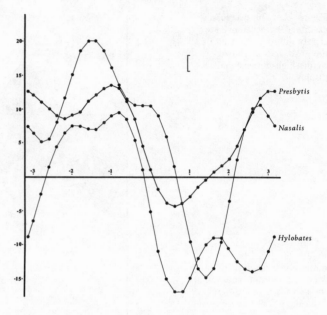

Figure 76.   This high-
dimensional display
demonstrates the clear
similarities of Kanapoi and
modern man.

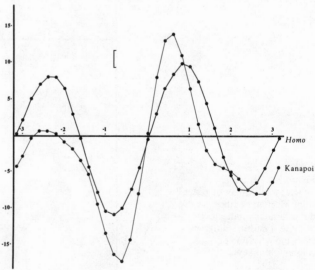

Overall Morphological Pattern

Visual Summation of Results from Individual Anatomical Regions

The weight of evidence in the previous chapters depends upon the analysis of a variety of dimensions of individual skeletal regions both of extant primates and of fossils. The separation of the analyses into discrete anatomical entities is, of course, deliberate, for it allows us to make assessments of each individual fossil remnant in an independent manner. We must be clear, however, that, once such assessments have emerged, some attempt should be made to ascertain the degree to which the various pieces of evidence can be added together. One way to do this is simply to add the information visually as is presented in table 4. In this way we can achieve two objectives. We can evaluate the results of the multivariate studies taken together as in the second column of the table. We can also compare these results with those assessments that are generally believed to be the case on the grounds of more traditional investigative methods (the first column of the table). A review of these latter assessments makes it clear that the fossils are generally thought to be very similar to man. The degree of closeness of this similarity varies from study to study, from worker to worker, and even, within the investigations of a given worker, from time to time. On the whole, however, it can be said that most believe that the fossils are similar to man and that when they deviate from the human condition, they do so in the direction of the African apes—man's closest genetic relatives and animals which, in their locomotor behavior, are terrestrial, knuckle-walking species.

The results of all of the work described in this book do not agree with this general conclusion. They suggest (second column of table 4) that these fossils are quite dissimilar to man or the African apes. Usually, indeed, they are so dissimilar from all hominoids as to appear unique, and in this sense, assuming that the forms of these postcranial elements are adaptive, they may be indicative of a unique form of locomotion. Similarities with extant primates are either in those characteristics that are so basic they can merely be said to be hominoid (i.e., similar to all apes and men) or in special characteristics that are mirrored in the orangutan to a degree. Presumably these latter indicate tentative functional similarities such as might be possessed by some intermediately sized arboreal form. We therefore may summarize this table by suggesting that these fossils did some of the things that arboreal animals do (for instance, acrobatic climbing and arboreal quadrupedality) but did enough other things (probably bipedality) that they are rendered unique in the frame of reference of present-day forms. There is certainly no evidence of functional similarity with the African apes, and no evidence of the same kind of bipedality as is characteristic of modern man. Such a conclusion is rather speculative although it is now based upon studies of all of the anatomical regions shown in table 4.

Is there any other way to add all this information together? Is there any other information that we can bring to bear upon this matter of uniqueness and diversity?

Tentative Computational Summation of Results

It would seem that one of the best ways to do this would be through multivariate statistical analysis as in the previous chapters. For example, sets of twenty measurements each could be taken upon each of twenty anatomical regions upon each

Uniqueness and Diversity in Human Evolution

TABLE 4

SUMMARY OF RESULTS OF MULTIVARIATE STUDIES OF
AUSTRALOPITHECINE[a] FRAGMENTS

	CONVENTIONAL WISDOM Resemblances on Man-African Ape Axis		CURRENT STUDY Resemblance	
	Homo	Pan-Gorilla	Orang-utan	Unique
SCAPULA[b]			●	
CLAVICLE[b]			●	
HUMERUS				●
METACARPAL				●
PHALANGES[c]			●	
PELVIS				●
TALUS			●	
TOE PHALANX				●

[a] In these studies the different species of *Australopithecus* are viewed as the single genus. Although there are certain differences within this fossil genus, the conglomerate seems no larger than the extant genera with which it is compared. Lumping is thus justified at this level of comparison.

[b] Morphometric investigation (Oxnard, 1968a, b, 1969b) not reported in detail in this study.

[c] Experimental stress analysis (Oxnard, 1973a) not reported in detail in this study.

specimen that is available for each of forty genera of extant primates. This would certainly be a major study, and for all except the larger computers it might well occur that the number of measurements taken upon each complete skeleton could be so great as to exceed the capacity of the computer for some of the matrix manipulations that would be required. Although for some investigators this computational constraint might well be serious, a far more practical reason why we are such a long way from such a study is the coordination that would be required to obtain such an extensive data set from so many anatomical regions of the same specimens of so many genera of extant primates.

And, of course, when it comes to the matter of trying to interpolate the various fossil fragments, they would have to be combined as though they presented information about the approximate means that might be obtained for the fossil genera. Inevitably one would have to pool data from fossil remnants that might well, in truth, belong to separate species, or perhaps even genera. However, this last limitation is not new and is with us all the time. The real problem is the immensity of the study that would be required to tackle all anatomical regions in an equivalent way.

Yet if the major study is not possible at this time, we must nevertheless try to summate the results of those studies that have been carried out in some manner or other. It is clearly possible to look at each individual investigation and summarize the results in the mind. However, there are other methods that we can use, and although these are not as good as direct summation by multivariate analysis of a total combined data set and do have obvious deficits that we shall examine, they may well give useful information that is difficult to see visually.

Thus, first, it is possible to summarize the information by "adding" the mean positions of generic groups for all the different anatomical regions once the discriminations of the generic groups have been determined for each region separately. This "addition" can be done using the high-dimensional display of Andrews (1972, 1973), but such a method is not as powerful as would be a display obtained from a multivariate analysis of the total data set. Secondly, we may try to obtain information from a wide series of anatomical regions by investigating a series of overall bodily dimensions for the entire group of living primates, and then, by careful investigation of form and proportion in such fossil fragments as exist, attempt to draw appropriate overall inferences (this second possibility is described in subsequent sections).

The first of these techniques means using information from the various regions already studied in earlier chapters (e.g., shoulder, pelvis, foot, arm, hand). This can be done by allowing the various significant canonical variates for each study to take the place of the independent unknowns in the sine-cosine equation outlined earlier.

To do this within each set of studies is entirely correct, for it is certainly known that the canonical axes are independent of one another—this is determined by the very nature of the procedure of canonical analysis. But to attempt this between the sets of canonical variates for each study is less rigorous, because there is bound to be some correlation between the sets of canonical axes for each anatomical region. If, however, these correlations are not overly high, then adding the various sets of canonical variates will give a picture that is not too far from the truth, and it will certainly allow us to see the complex multidimensional structure.

That the correlations between the different major anatomical regions will probably not be high is shown by the results of studies in which, among primate genera, groups of dimensions of, for instance, the forelimb are but poorly correlated with those of the hindlimb. Thus out of forty-eight sets of correlations between eight forelimb dimensions and six hindlimb dimensions, only eight lie above the 0.5 level, and eighteen, almost half, are negative and near zero. In comparison, of the correlations between sets of measurements within the forelimb, only four out of twenty-eight are negative (and near zero), while thirteen lie above the 0.5 level; similarly for the hindlimb, of fifteen correlations within the hindlimb only four are negative (and are in fact close to zero), while six are above the 0.5 level.

Even within major bodily regions, anatomical entities rather distant from one another (e.g., measures of the shoulder as compared with the hand, measures of the hip as compared with the foot) are rather poorly correlated with one another, usually below the 0.5 level; measurements from contiguous anatomical regions display correlations that are considerably higher, usually above the 0.5 level.

A second piece of information that suggests that the various canonical variates from the different studies are likely to be rather poorly correlated with one another (i.e., probably contain rather different information) is the fact that the picture shown by canonical analysis of one region (e.g., the hindlimb) is quite different from that

shown by the canonical analysis of another region (e.g., the forelimb), as can be seen
in figure 77. A similar comparison of studies in the scapula and pelvis appears to
provide further confirmatory evidence (fig. 78).

In any case, we cannot carry out the more elaborate and correct procedures; therefore, anything that the somewhat looser method outlined here can provide, may, if

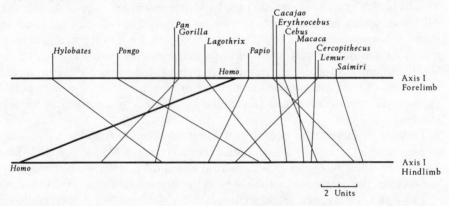

Figure 77. A comparison of the information contained within canonical axes one of the individual
studies on the eight dimensions of the forelimb and the six dimensions of the hindlimb. That the
arrangement of the genera is completely different in these two investigations is shown by the lack
of concordance between their positions in each axis. Some of the lines connecting a genus in each
study are vertical, indicating similar placement; others are markedly oblique, indicating different
placement. Vertical and oblique lines occuring at almost every location indicate that there is little
association between the relative placement of genera in the two studies. Compare figures 24 and 27
of chapter 2, which, in contrast, show marked associations between pairs of studies.

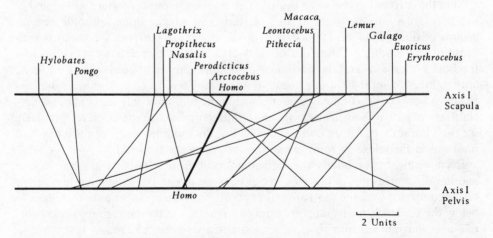

Figure 78. A comparison similar to that in the previous figure for the arrangement of genera in the
other two studies, one on the seventeen dimensions of the scapula, the other on the nine dimensions
of the pelvis. The picture is entirely similar to that in figure 77, indicating almost no association
between the results of the two investigations. This compares with the marked associations that are
shown in figures 24 and 27, chapter 2.

viewed with appropriate caution, give considerable useful information. Certainly in such a multidimensional situation some help in terms of statistical reduction and display is absolutely necessary.

Accordingly, the results from some of the analyses of earlier chapters (those of the pelvis, talus, and humerus) have been compounded using the method of Andrews (1972), and this provides a set of functions that represent a fifteen-dimensional data-space. The sine-cosine curves that can be derived from each genus are extremely complicated with a tortuosity greater than in any previous chapter. Rather do they resemble the complicated curves shown in Oxnard (1973c), where a twenty-three-dimensional picture is shown for overall bodily proportions of some Prosimii. Figures 79 and 80 provide the results of this additional display for certain extant genera of the remaining primates, and they suggest rather clearly that the three great apes are very different from one another, although, as we would expect, the orangutan is considerably further from the gorilla and chimpanzee than these latter are from each other (fig. 79a). When we also compare man we find (fig. 79b) that he lies even more distantly from the African great apes. This display helps us to appreciate that the four large hominoids differ markedly from one another, although *Pan* and *Gorilla* do form a subgrouping. This is a result that scarcely adds to what we know from simple observation.

When we include the composite information from the fossil specimens with this admittedly somewhat speculative procedure, we find (fig. 80a) that the fossil is also very different from the African great apes. But when we compare the position of the fossil composite with those of man and the orangutan, we see that, with little exception indeed, it lies almost entirely intermediate between these two living forms; the curve for the fossil composite lies, to a quite remarkable degree, within the envelope formed by the positions of the curves for man and the orangutan (fig. 80b).

This relationship cannot be of taxonomic or genetic import. If the discussions of the previous chapters have any reality, this result, notwithstanding the less rigorous nature of the method that has provided it, presumably reflects morphological adaptation relating to arboreal activities characteristic of acrobatic animals as well as terrestrial abilities somewhat reminiscent of bipedal man. As such it stands as an excellent summary of the previous findings.

Overall Sizes and Proportions of Primates

A second way to attempt some overall view of the morphology of the entire class of primates is to assess the relationships obtained among the living genera when data about their overall form and proportions are appropriately analyzed. Fossil information again cannot be directly interpolated into such studies without totally unwarranted assumptions, full of enormous error about possible overall form and proportions in species for which we have only fragmentary information. However, as we shall see later, once we understand the relationships given for extant genera by such an analysis, there are indirect ways of fitting in fossils that do not depend upon tentative estimations or frankly speculative guesses about the sizes of parts that are missing.

Figure 79. These plots indicate the high-dimensional curves, derived as explained in the text, for the pelvis-talus-humerus combination. The upper frame (a) shows that the plot for the orangutan differs markedly from that for the envelope of the African apes which itself shows separation of these two forms. The lower frame (b) shows that man differs even more from the African apes. Cross diagram synthesis of the curves for man and the orangutan indicate that these two also are very different. This method of display is, as explained in the text, somewhat less rigorous than when used for a single study as elsewhere in this book.

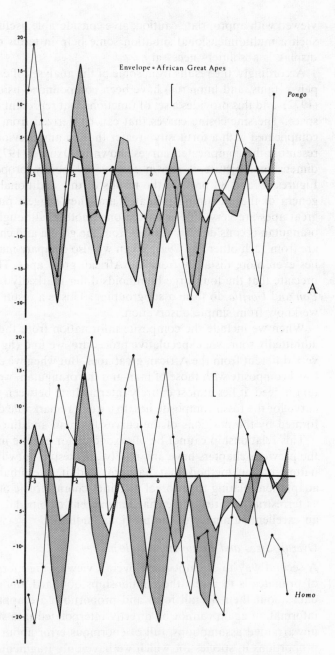

In this series of investigations we have returned to the studies of Professor A. H. Schultz of Zurich. His series of measurements, compounded into indices, of the bodily proportions of a wide range of primates are virtually unrivalled save by the data of Erikson (1963), and for technical reasons it is not feasible to add the two data sets together.

Figure 80.   As with figure 79, these plots indicate high-dimensional separations for the pelvis-talus-humerus combination. The upper frame (a) shows that the fossil combination (of specimens from Kromdraii, Olduvai, and Sterkfontein) is markedly different from the envelope characterizing the African apes.

The lower plot (b), however, shows that the fossil is almost entirely intermediate between the curves for man and the orangutan. Only a single peak at the right hand end of the plot indicates elements of the fossil that depart from an intermediate location. The same qualifications apply to this figure as the last.

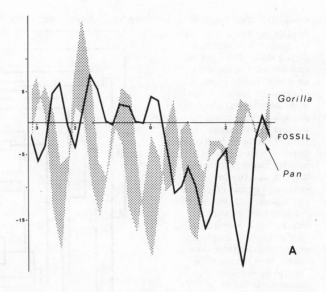

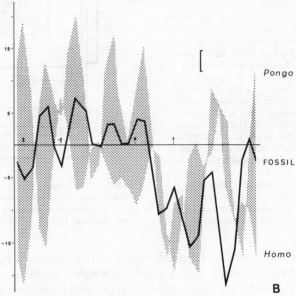

Some features of this data set have already been provided by Ashton, Flinn, Oxnard, and Spence (1974) who suggest that when all bodily dimensions are compounded, the primary relationships of the various primate genera resemble those most closely given by current taxonomy. Further studies of this data set are showing that in certain aspects the results depart from current taxonomy but that, when they do, they do so in ways for which explanations are rather easily evident (figs. 81 and 82).

Figure 81. This figure displays the dendrogram of the relationships obtained from a study of the generalized distances of the data on overall bodily proportions of primates. Some relationships are relatively clear—for example, (a) the grouping together of the indriids, one form of "vertical clingers and leapers," with *Lemur* as a possible incipient form close to them, and (b) the grouping together of *Tarsius* and *Galago*, a second set of "vertical clingers and leapers," with *Microcebus*, another possible incipient or intermediate form close to them. These relationships seem to be obviously locomotor in origin in addition to whatever other information they may display. However some relationships are not well displayed by this method. Thus while *Pongo*, *Hylobates*, and *Symphalangus* are clearly preferentially linked, the two African apes appear to be more closely allied to the Old World monkeys. This is of course the case in the dendrogram, but what the dendrogram is not capable of showing is the fact that the conjunction between the Asiatic apes and most of the other primates are through links with the African apes.

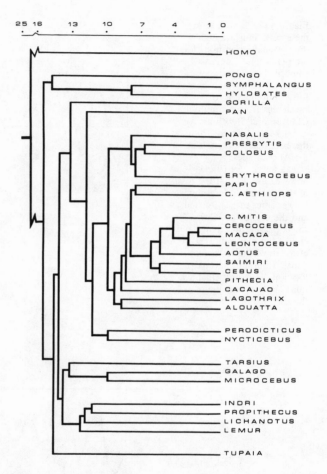

Relationships among the various genera are better seen from the minimum spanning tree (see next figure).

For instance, the analysis of bodily proportions by multivariate statistics departs from taxonomy to the degree to which it places forms such as *Galago* and *Tarsius* together (figs. 81 and 82). However, these forms, taxonomically grounded in different infraorders of the Prosimii, are so highly adapted for a similar locomotor pattern—one type of vertical clinging and leaping (Oxnard 1973c, and Stern and Oxnard 1973) that a resemblance is entirely accountable on this score.

In other instances current taxonomy is departed from in other ways. For example the New World monkeys are divided by these analyses into three groups, the more acrobatic prehensile-tailed Cebidae (*Ateles*, *Lagothrix*, *Brachyteles*, and *Alouatta*), the apparently more generalized genera (*Aotus*, *Callicebus*, *Pithecia*, *Cacajao*, *Chiropotes*, *Callimico*, *Callithrix*, and *Leontocebus*), and the intermediate genera

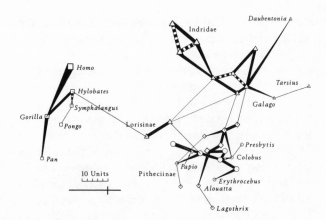

Figure 82. The minimum spanning tree of generalized distance connections superimposed upon the directions provided by the bivariate plot of the first two canonical axes for the overall bodily proportions of the primates. The different thicknesses of the lines are attempts to indicate the relationships of groups within a third dimension in a situation where there are spurious overlaps. The dotted lines indicate points where there are several connections that are all close to the minimum one. However, because of the high-dimensionality of the situation the only distances between groups in the diagram that are correct are those represented by solid lines.

The major separation, Hominoidea and Prosimii, is clear. The apparent overlap between the New and Old World monkeys is shown by the connections to be spurious—these groups are rather well differentiated from one another. Within the Hominoidea, the relationship of the orangutan is with the lesser apes, the relationship of man is with the African apes. Within the Prosimii, the Indridae and the Lorisinae are well demarcated. Other outlying genera are *Daubentonia* and *Tarsius*. There are good separations within the New World monkeys of the prehensile-tailed species and of the Pitheciinae; within the Old World monkeys the colobines are distinguished as outliers, as also are the terrestrial genera *Papio* and *Erythrocebus*.

(*Cebus*, also prehensile-tailed, and *Saimiri*). This differs somewhat from the currently recognized groups (the family Callithrichidae, and the subfamilies Aotinae, Pitheciinae, and Callimiconinae). In this case, however, the current taxonomy is not strongly agreed upon by all authorities, and the grouping presented here is probably as good a taxonomy of New World monkeys as is the currently accepted one. (This is not, of course, any argument for changing the current taxonomy of the group).

Finally, in similar vein, the association here (figs. 81 and 82) of orangutans with gibbons rather than with the African great apes (compare current taxonomy which allies the orangutan with the African great apes in the family Ponginae) is probably also as entirely a realistic taxonomic association, if not considerably more so, than is current taxonomy.

Certainly all of these departures from the current taxonomy are supported by data that can be derived from molecular studies (e.g., see Goodman 1973 for review). These particular sets of associations therefore cannot be viewed as other than reasonable, some of the departures from current taxonomy being clearly due to inclusion of information about parallel functional adaptations, others representing areas in which current taxonomy is itself somewhat questionable.

But when we come to look at subsets of the data on overall bodily proportions, other information is apparent. Thus studies on the dimensions of the forelimb alone show remarkable similarity with those already determined from the analyses of the much more restricted region of the shoulder (fig. 24, chap. 2). Similarly, studies on overall dimensions of the hindlimb alone show relationships among the various extant primate genera that are rather closely parallelled by the more restricted study on the smaller anatomical region of the pelvis and also to some degree by the more restricted study of the talus (fig. 27, chap. 2).

Multivariate statistical analyses of the overall bodily proportions of the head and trunk are not, however aligned with either of these previous sets of functional groupings. In this anatomical region, most of the information seems to relate to the discrimination of various individual primate genera. Thus the dimensions of the head and trunk taken together provide a series of separations that on the one hand group all of those species that are relatively generalized in their morphological features (e.g., among the Prosimii, genera such as *Lemur, Hapalemur,* and *Lepilemur;* among the New World monkeys, genera such as *Aotus, Callicebus, Callithrix,* and *Leontocebus;* and among the Old World monkeys, genera such as *Cercocebus, Macaca,* and *Cercopithecus*). At the same time, however, these data tend to separate out those various genera which are known from other data to be aberrant in some form or other. These include among the Prosimii the genus *Daubentonia* (the so-called aberrant aye-aye) and the genus *Tupaia* (a genus which is likely not to be a primate at all). Outlying forms among the New World monkeys are the highly specialized genus *Alouatta* which although it has affinities to prehensile-tailed forms nevertheless seems to be relatively unique, and the genera *Cacajao* and *Pithecia* which also display some considerable differences from the bulk of New World forms. Among the Old World species, one unique form is the exceedingly terrestrial genus *Erythrocebus,* which is markedly separated from the other cercopithecines; others are the also markedly terrestrial genera *Papio* and *Mandrillus,* likewise rather uniquely separated from the main bulk of cercopithecine forms.

It is the addition of this information about specific genera in the discussions of the trunk and head to the information about locomotor adaptation in the limbs that produces, in the total study of all dimensions combined, a picture that is close to that presented by current taxonomy. There is thus a great deal more power in these overall dimensions *when they are combined* than has ever been previously suspected. How useful it would be to have equivalent information for a fossil.

How can we utilize this information in the study of incomplete fossils? Clearly it is unreasonable in the case of the *Australopithecus* to attempt to reconstruct these overall bodily proportions. This can be done for fossils with some success when the remains are as extensive as, for instance, are known for *Notharctus* (fig. 83). In this case a number of overall bodily proportions for *Notharctus* have been derived, and it has been possible to interpolate them into appropriate multivariate statistical analyses (Oxnard 1973c). But it is inconceivable, when the gaps are as large as those for *Australopithecus* (for no single individual of which are there more than a few fragments of localized anatomical areas—e.g., pelvis, lower vertebral column, upper femur—STS14), that reconstruction of major limb proportions and overall bodily dimensions can be reasonably attempted. Although this has been done by some workers (e.g., reconstruction of limb length and overall stature for *Australopithecus*), nevertheless, in the main, these reconstructions have been made assuming that proportions for man are a reasonable model and that, therefore, regressions based upon statistics for man can be utilized in the reconstruction. As we shall see, there is considerable information that suggests that this is entirely unwarranted, and if regressions based upon some other species had been used, quite different estimations would have resulted. One of the earliest intimations of this has been provided by McHenry in his comparison of the talus and lower end of humerus (1972) in which the disproportion between these elements in man and *Australopithecus* is clear, although no comparisons are provided with any ape. It has also been noted by Leakey (1973a), who reports that "preliminary indications point to a relatively short lower limb and a longer forelimb" for *Australopithecus*, and that this raises the possibility as to "whether the australopithecine pattern of bipedal adaptation really reflects a transitional phase" toward that of man.

Relative Sizes of Fragments of Anatomical Regions

However, although we cannot utilize at this stage these kinds of estimations of overall bodily proportions, what we can do is to investigate the relative sizes of the fragments that we do have, and we can do it in a comparative way that provides us with a major impression about overall form and proportions.

Let us be clear that we are not here talking about detailed facets of shape of the individual specimens, nor are we discussing reconstruction of the lengths of bones. Rather, we are speaking of the overall impression that may be gained by study—in this case, of a rather classical type—of cross-comparisons of the sizes of the actual existing fragments.

Thus if we confine our attention in the first instance to the forelimb, we can compare the sizes of the individual anatomical parts of the various fossils against the sizes of equivalent parts, obtained from actual individual complete skeletons of a man, a chimpanzee, and an orangutan. The particular extant subjects have been rather carefully chosen to be the same absolute size for the most proximal articulation of the forelimb, the scapular articular surface: the glenoid cavity. The dimensions of the fossil are those given by Robinson (1972) although we have chosen to draw the fossil as a fractured specimen, as it actually is, rather than try the reconstruction attempted by Robinson. Robinson's reconstruction is based upon the specimen, whereas any reconstruction that we might make would have to be based upon drawings. Because the anterior and posterior margins of the fractured glenoid cavity—we are here indebted to Campbell for the drawing used in this study—cannot be matched into smoothly flowing lines such as must have been possessed by the fossil during life, we have not attempted reconstruction.

This diagram of the glenoid cavity thus becomes the reference, and the remaining glenoid cavities in figure 84 have been obtained by deliberately choosing *Homo* 89 from the University of Chicago, Department of Anatomy, and *Pan* 27752 and *Pongo* 33533 both from the Field Museum of Natural History, Chicago, because they are very nearly the identical size. It is likely that the dimensions of these

Figure 83. Sketch of
skeleton of *Notharctus* based
upon Gregory (1920) showing
relatively minor deficiencies
(white) in the fossil materials
for a single known specimen.

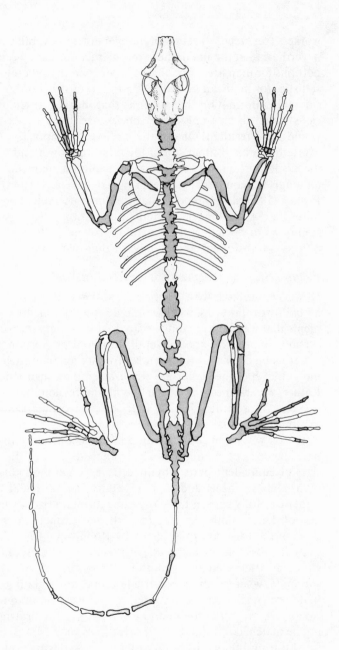

articular surfaces are related at least in part and in a general way, to the forces acting across the joints during normal functions. Hence these four articular surfaces
may have had generally similar loads during life placed upon them, although, of
course, since the samples come from different species, the nature of the loading and
the range of movement involved would presumably have been quite different.

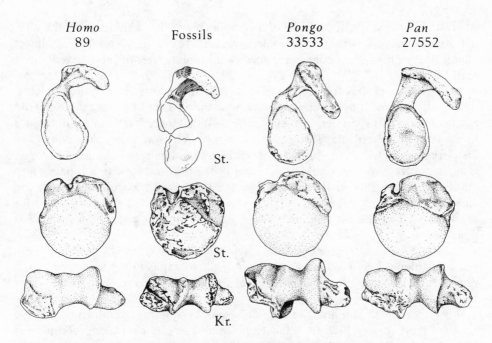

Figure 84. Articular surfaces from forelimb elements for a particular specimen of *Homo, Pongo,* and *Pan,* chosen so as to have near identical dimensions for the glenoid fossa. Compared with these are fossil fragments from Sterkfontein (2) and Kromdraai drawn to the same scale. The overall similarities in size of all four sets are obvious. Although the extant forms figured here are not representative of the average absolute sizes for their own genera, they are representative in proportions. The figure thus confirms the general relative similarity in forelimb articular proportions in extant species and fossils.

Once this has been done, it is possible to make correctly scaled drawings of various other anatomical parts from these same extant specimens. Each comparison will indicate visually how the proportions of one anatomical part to another differ in the extant forms. Further comparison with the various fossil fragments that are available gives an impression of how such proportions may have differed in the fossil and of how these differences relate to those between the extant forms. However, because the fossil parts do not come from a single animal or subject, the nature of the fossil comparison will be less precise.

The next anatomical region that we may examine is the articular surface on the upper end of the humerus. Figure 84 shows, that, as with the scapular glenoid cavity, these humeral parts are almost exactly the same size in the extant forms. We do not find this unexpected, of course, because this surface is after all, the functional mirror of the glenoid cavity in the previous example. It is rather comforting, therefore, to discover that when a different fossil find of the upper end of the humerus is included in this comparison (specimen number STS 7), a generally similar picture is presented, although it is true that this specimen is slightly smaller.

At least for both scapular and humeral components of the shoulder joint, the relative sizes among all the higher premates figured here are, on a visual level, almost identical, even though, of course, the means of absolute sizes of these genera differ markedly.

We can proceed thus for the articular surface at the lower end of the humerus. Basing our judgment again upon the nearly identical sizes of the scapular glenoid cavity in the extant specimens, we can see (fig. 84) that there is rather little difference in the sizes of the articular surfaces at the elbow. When we interpolate one of the fossil fragments (specimen number TM 1517) in the same way (and this, of course, is again not from the same subject as either of the previous fossils), it is once more gratifying to find that in its general size, the fossil is not too different from the various extant forms, being only slightly smaller. Repetition of this comparison for the front view of the distal humerus (not figured) entirely corroborates what has gone before.

Accurately sized casts of more distal fragments of the forelimb in these fossils are not available to me although examination of photographs in the literature (ulnar, radial, carpal, metacarpal, and phalangeal fragments) tends to confirm the views provided above.

It is important to note that we are not here discussing the overall lengths of the bones of these species, but rather the actual overall sizes of equivalent specimens at joints (e.g., see Oxnard 1973d). The articular surfaces are of similar size in the specimens used, thus suggesting perhaps approximately similar degrees, if not types, of loading. The known lengths of the bony levers in the different extant specimens differ markedly in relation to the totally different way in which the bones are loaded in the different specimens. But as we have a reasonable total length for the fossils for only a very few anatomical regions (clavicle, tibia, and fibula), this information can scarcely be utilized.

When however, we attempt a similar series of comparisons for those same extant specimens but upon the hindlimb elements, a series of entirely different findings results. Let us first confine our attention to the hip articular surfaces that are available. The comparison of the auricular articular facet, the acetabulum, and the femoral head is given in figure 85. In the case of each of the extant specimens—Homo 89, Pan 27752, and Pongo 33533—the pelvic and femoral articular surfaces of the chimpanzee are smaller than those of man; those of the orangutan are smaller still, all in comparison, of course, with equal-sized glenoid articular surfaces.

Comparison with those fossils that are available show that they too are smaller than man and approximate in size to the apes (specimen numbers STS 14 and SK 82, fig. 85).

As we pass down the hindlimb, these differences become rather more marked. For instance, at the knee joint (fig. 86), the lower end of the femur and upper end of the tibia are even smaller in our chimpanzee as compared with our human subject than was the case at the hip. And our orangutan specimen can now be readily distinquished, on a size basis, even from the chimpanzee, because of further reduction of these parts. Thus a distinct size spectrum exists at the knee from the human specimen through the chimpanzee to the orangutan.

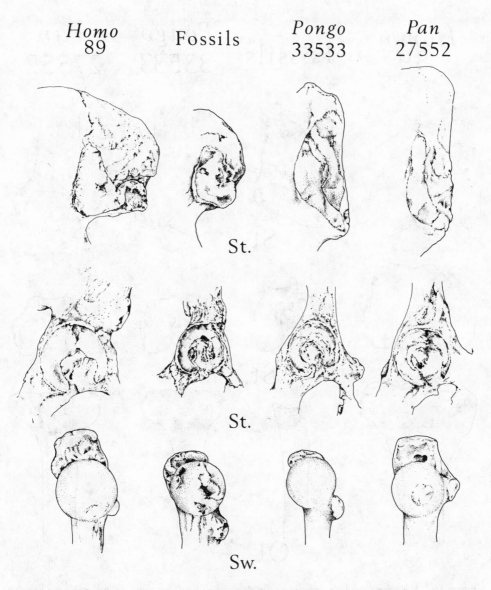

Figure 85. This shows some proximal articular surfaces of the hindlimb from the same extant specimens as figure 84. It is immediately apparent that, irrespective of different shapes, and irrespective of similar relative articular proportions in the forelimb, the articular proportions of the proximal elements (auricular facet, acetabular fossa, and femoral head) of the hindlimb are far from similar. Man is enlarged compared to the extant apes. The fossils (from Sterkfontein and Swartkrans) show relative reductions comparable to those of the two apes.

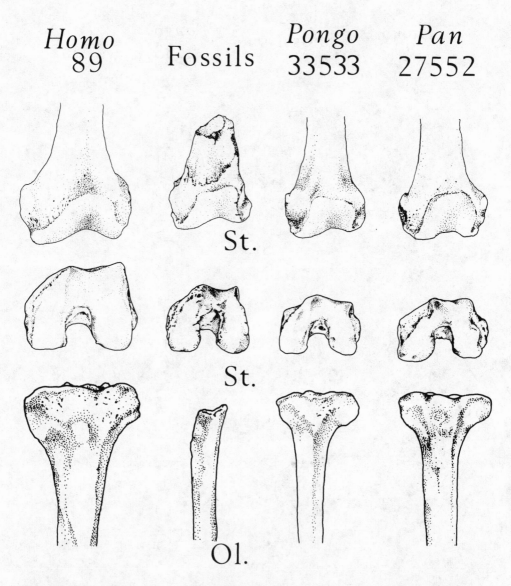

Figure 86.   This shows similar comparisons for elements of hindlimb articular surfaces at the knee. Again there is reduction of elements in the two pongid specimens as compared with the human subject. But, in comparison to the hip, the reduction for the orangutan is somewhat greater than for the chimpanzee. Fossil specimens (from Sterkfontein and Olduvai) show reductions that are markedly different from man, and are greater even than those of the chimpanzee. The relative proportions at the knee, as compared with those at the forelimb, are almost as reduced for the fossil as are those of the orangutan.

The position of the various fossils is thus of extreme interest, and, again with the caveat that we are not comparing the same animal but with the *hope* that perhaps these particular specimens are at least somewhat representative, we can see that the australopithecines have knee fragments much smaller than man as compared with the forelimb fragments: we can see that the fossil surpasses even the chimpanzee in terms of the relative reduction of these fragments; the degree of reduction is every bit as great as that of the orangutan (fig. 86).

When we examine those articular surfaces that are even more distal, the lower tibial and upper talar surfaces, and the dorsal view of the navicular bone from an articulated foot (see also Oxnard 1973d), we can see yet further continuation of this trend. Size for size, the chimpanzee is reduced even more compared with the human specimen and the orangutan is very greatly reduced, to an extent that is greatest the more distally that we progress (fig. 87). The fossil, size for size, resembles most clearly the orangutan (fig. 87).

Finally we may view all of these differences in the hindlimb together. The pattern in the various fossil fragments displays a gradually increasing proximo-distal reduction which differs from that shown by the chimpanzee; rather it is with the orangutan that the gradually increasing reduction of hindlimb articular elements is most similar. Even here, however, there are differences, for though the foot elements show the most marked reduction characteristic of the orangutan, the hip elements tend to show a somewhat lesser reduction more like that in the chimpanzee. Proportionate resemblances to man there are not.

This information gives us some idea of the relative changes of the sizes of different anatomical parts as we progress over the skeleton in these different forms. It suggests rather strongly that the differences observable between man, the African apes, and the orangutan and based upon relative proportions primarily of articular surfaces, may well be replicated in these fossils in such a way that, irrespective of the actual absolute lengths of the elements which we cannot at this stage know with any confidence, their relative proportions may fall rather close to those of orangutans. As this information is entirely confined to limb fragments, it seems clear that we are looking at degrees of resemblance that have a greater functional than taxonomic import. Once again we are led to look at locomotion in the orangutan for at least some of the biomechanical modes that may be relevant. It is also clear, however, that these fragments possess features that are not only different from the orangutan, but that differ also from all of the other forms; we therefore have to be prepared to postulate additional unique features for their locomotion.

What functional meaning is it possible to associate with the foregoing information? Starting from a premise of equivalently sized elements of the upper limb in particular specimens of man, the great apes, and the fossils, we find that, in the hindlimb, man presents a marked increase in size compared with the chimpanzee. This presumably correlates with the emphasis in man on the major forces of weight-bearing being confined to the hindlimb, whereas in the chimpanzee they are shared between both sets of limbs because of the habitually quadrupedal gait of this species.

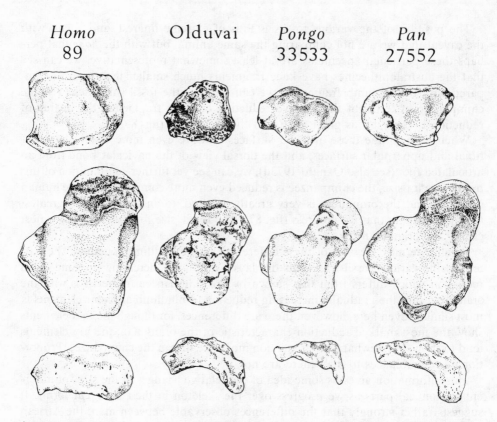

Figure 87. This shows the final set of comparisons for articular surfaces from the same extant specimens at the ankle and foot. The reducing trend that was evident in the hip and knee is further continued at the ankle so that, although the chimpanzee is considerably smaller than man in proportions at the ankle and foot relative to the forelimb, the reductions are much greater for the orangutan. Examination of the fossil from Olduvai indicates that the fossil is reduced at least as much as the orangutan, if not more so, compared with the equivalence in size of forelimb elements portrayed by figure 84. Again it must be emphasized that these series of figures relate to cross-comparisons between the two limbs in the various species; the actual mean absolute sizes for the genera are not represented here.

In contrast, we find that the orangutan presents a marked decrease in size of hind-limb elements as compared with the chimpanzee. Perhaps we can take this as cor-relating with, in the orangutan, an emphasis in weight-bearing by the forelimb which, as compared with the chimpanzee, might be entirely expected in an animal that spends a great deal of time in forelimb acrobatic endeavors (this is not to say that the hindlimb does not participate—it only notes that in the orangutan the emphasis is rather more toward the forelimb than it is in the chimpanzee).

It is of interest, therefore, to discover that the fossil, unless chance has played us the very dirty trick of providing forelimb fragments from large specimens, and hindlimb fragments from small specimens,[1] resembles generally the orangutan in the degree of emphasis upon relative size in the postcranial articulations.

Yet, at the same time, we recognize that the fossil is by no means identical to the orangutan. For instance, the reduction that is so marked in both the orangutan and the fossil in the foot is not nearly so evident in the hip. Here the fossil is bigger than it should be were it simply mimicking the orangutan. This degree of emphasis on the pelvic end of the limb may be especially important in our assessment of the uniqueness of the fossil locomotor mode. For although reduction in the foot might, as in the orangutan, betoken abilities along the lines of acrobatic–arboreal-climbing activities, increase in the pelvic area might, as in man, relate to a biomechanically better and relatively more habitual type of bipedality than any of the great apes are capable of. At the same time, there are no really close resemblances with man so that the bipedality of the fossil, if present, must have differed markedly from that of modern man. It is difficult at this stage to know how the ecological information can be taken into account, and difficult also to visualize the "surprises" that future fossil finds may hold for us.

Australopithecines, Man, and the Evolution of Locomotor Function

The foregoing results represent the end products of a fairly large number of studies involving many different anatomical regions spread over the whole of the post-cranium (shoulder, elbow, hand, pelvis, hip, knee, ankle, and foot). They are the resultant of a series of multivariate statistical procedures that are capable of revealing facets of information that are hidden from more usual visual inspection. At the same time they are consonant with a series of studies that do utilize visual inspection as the primary method but that view the fragments in a manner not usually attempted. The results are generally corroborative of one another when seen separately and independently. They continue to provide the same story when attempts are made to "add" them together (albeit by techniques that can, at this stage, only be somewhat speculative) in order to try to obtain an overall view.

That result is the following: in terms of morphology, the various australopithecines, viewed as a single group at the higher taxonomic level, are generally more similar to one another than any individual specimen is to any living primate. They are uniquely different from any living form to a degree comparable at least to the differences among living genera. The manner in which they are similar to living apes and man is either such as is applicable to all living apes and man, or such that displays special morphological resemblances to a particular ape, the orangutan. In few ways are they more similar to either man or the African apes than these latter are to each other.

The meaning of this result is far more difficult to disentangle. It undoubtedly does not rest upon genetic propinquity, as it is surely clear that genetically, man, the African apes, and presumably the australopithecines are all closer to one another

1. And in the hindlimb, the smaller the more distal the fragment.

than are any of them to other living hominoids. Because of the interpretations that
can be made from functional anatomy, it is presumably far more likely that the
patterns of uniqueness and diversity that have been uncovered relate to the evolution of locomotion.

Accordingly, the more speculative deductions that these results speak to are the
following. Just as they display uniqueness in morphology, the australopithecines
may likewise have been functionally unique from all living hominoids. This surely
means that they were not arboreal in the different manners of the Asiatic apes;
they were not terrestrial in the knuckle-walking mode of the African apes; they were
not solely bipedal like man. They therefore displayed either a totally new and unknown manner of locomotion which would be totally unique and which we will
judge rather unlikely; or they displayed such a mixture of locomotor abilities, therefore anatomical adaptations, and therefore bony morphologies, as to be rendered
unique through being a mixture of different intermediates.

The modes in which they are similar to man indicate propensities for a type of
bipedality. But the ways in which they are similar to orangutans may well also indicate abilities for quadrupedal movement, presumably in an arboreal environment
and with a degree of acrobatic climbing abilities such as are reminiscent of all apes
when they are in an arboreal milieu (or, just possibly, a clifflike terrain). Let us be
clear that the nature of these suggestions is not merely an extension of human
capabilities. Man can climb using methods like those of apes rather than like any
monkey. But his abilities to live in a climbing setting are almost nonexistent. Likewise, man can—and does, when an infant—move on all four limbs, but in fact his
quadrupedal facility is zero. Man's true abilities in both these directions cannot
be said by anyone to be other than a total liability in any conceivable life framework
where they are required. Such was presumably not the case with *Australopithecus*,
although we are not able to disentangle the time relationships of these abilities.
Australopithecus may display these morphologies because it had both sets of abilities
or because, while performing the one as a new acquisition, it had not yet lost the
hallmarks of the other, older, mode.

Finally, we may make the further speculation as to what all this may mean for
human evolution and the position of the australopithecines within it. It is still possible that these results are consonant with the view that australopithecines are close
to the pathway of the evolution of bipedality as expressed in the evolution of man.
This is the standard view. If this is true however, it suggests that bipedality arose by
some markedly complicated process involving genetic intermediates that are very
much unlike man and that this all occurred in an extremely short period of time.
These results are actually rather more consonant with the simpler idea that human
bipedality was not the only experiment in this functional direction. The australopithecines may well be displaying for us another experiment in bipedality—one that
failed. Further, they may even be providing for us evidence of *several* failed experiments, for we have no guarantee that they do not comprise several evolutionary
lines. The reasons why this may have happened are speculative in the extreme and
at this point scarcely deserve our attention.

Three other pieces of evidence support these ideas and are independent of those assembled in this study. One comes from the recent finding at East Rudolf by Richard Leakey (Leakey and Wood 1973) of a talus that was found lying between layers that have been reliably dated at 1.57 ± 0.00 and 2.61 ± 0.26 million years by argon isotope techniques. This specimen is thus as old as, if not appreciably older than, the australopithecine specimen from Olduvai (the dating of Kromdraai is problematical). Yet this new specimen is much larger than the australopithecine specimens and appears to be much more like man as far as can be judged from the published picture and descriptions. Further description and examination using canonical analysis by Wood (1974) confirms that it is indeed very similar to modern man and is thus unlike the australopithecine specimens. Unless evolution took the talus through a stage where it was much like man (as at East Rudolf), then through a stage where it was uniquely different from man (as at Olduvai and possibly Kromdraai), and back again to a stage like man (modern man), then the australopithecine fossils had to have been unrelated to any direct human line.

A second comes from the recent finding by Richard Leakey (1973b) of a skull dated as at least two million and perhaps almost three million years and with an endocranial volume of at least 700 ccs and possibly 800 ccs or even more. Such a find—if the accurate dating can be certainly pinned to the fossil rather than to the layer in which it was found (and this seems likely), and if the volume can be independently confirmed (it certainly looks reasonable from the casts of the reconstruction that I was allowed to examine[2]), makes it much less likely that the australopithecines, the youngest of which is less than one million years and the endocranial volume of which is just over 400 ccs—the same as a large gorilla disregarding overall bodily size—are on any direct line of human ancestry. Unless cranial capacity went from large to small then large again, the new find makes it clear that the australopithecines had to have been off the main stream of man's development.

The third piece of evidence comes from an even earlier find, the fragment of arm bone perhaps four million years old from Kanapoi. This has already been shown by Patterson and Howells (1967) to be very similar to that of modern man, and some of the demonstrations in this book clearly support that contention (chapter 5). Again, unless arm bones evolved through a cycle of being more like man at an earlier, Kanapoi, stage, less like him at later australopithecine stage, and more like him again at a much later human stage, then the australopithecines had to have lain on an evolutionary side branch.

And these pieces of evidence must also mean that perhaps as long as five million years ago (and the possibility is not lost that future finds may place this yet further back in time) there may well have been creatures living that were generally similar to *Homo erectus* and therefore classifiable as man in a way that we must deny to any australopithecine (whether named "*H. habilis*," "*H. africanus*," or whatever else). The removal of the different members of this relatively small-brained, curiously unique genus *Australopithecus* into one or more parallel side lines away from a direct

---

2. Although one expects the estimate to be subsequently improved.

link with man would make it easier for the mind to accept the findings displayed in this book. But whatever the explanation may be, the findings in this book are real; the australopithecine fossils are much less like man than is currently recognized. These results must be accounted for in whatever conclusions are drawn concerning the evolution of man.

This much earlier existence of individuals somewhat like *Homo* (in the very few features available for comparison) leads inexorably to a further much broader possibility. That is: the length of time that would have been available for the psycho-social evolution of the genus *Homo* may well have been almost an order of magnitude greater than that presently believed to have been operative. *Homo erectus* is generally thought to be less than one million years old, although Howells (1973b) percipiently argues for a longer period. Knowing as we do the enormously greater speed of psycho-social evolution as compared with the slow rate of biological evolution, then a longer absolute time span of, say, five million years, may have allowed an even greater amount of relative evolutionary time for the evolution of the behavioral, cultural, and intellectual qualities that stamp man as unique from any animal. Thus it may be that the exceedingly cruel yet sophisticated nature of human aggression towards his own species is less dependent upon a prior animal base and more dependent upon the relatively long psycho-social evolutionary time span that we may now suppose to have existed. The extreme intricacies of human culture may likewise reflect somewhat less upon a previous basic behavior as pertaining to a nonhuman primate and somewhat more upon this extra period of evolutionary time than has previously been supposed by many workers. Human intellectual prowess may have had as much as ten times more time to increase from a level comparable to that of some nonhuman stage of development than we have previously realized.

It may well be worth making a search for some estimate of the ratio of the speeds, or more probably of the accelerations, of psycho-social as compared with biological evolution. This would allow us to see by just how much the five million or even more years that may have elapsed since the appearance of something somewhat similar to *Homo erectus* must be multiplied in order to give an equivalent to our usual view of biological evolutionary time. The results might be quite surprising.

One would expect that Leakey (and others) will find more evidence of earlier forms that are considerably more human in all respects. These would, on present evidence, be likely to be more like man than the later australopithecines (even though these latter must also have existed at earlier periods in time). Certainly, the numbers of fossil remnants that are now appearing through the work of many investigators in Africa would yield a great deal of information if compared with living species by the methods used in this essay. We may well have to accept that human bipedality is far older than previously guessed and that australopithecine locomotion included one or more unsuccessful parallel experiments. We may well have to accept that it is rather unlikely that any of the australopithecines, including "*Homo habilis*" and "*Homo africanus*," can have had any direct phylogenetic link with the genus *Homo*.

All of this makes us wonder about the usual presentation of human evolution in encyclopedias and popular publications, where not only are the australopithecines described as being of known bodily size and shape, but where, in addition, such characteristics as bipedality, tool-using, and even facial features are happily reconstructed.

One final comment is in order. Such similarities as have been found among many of the fossil fragments examined in this work and the equivalent parts of the orangutan do not suggest that the ancestors of the orangutan played any part in human evolution. The morphological similarities can surely only betoken some functional parallels for the bony shapes involved. These new comparisons may be supplying us with important clues about the australopithecine fossils, the locomotion of which, if we remove them from a close evolutionary link with man, we now have to investigate as of a terminal group fascinating in their own right.

Note Added in Proof

Since making the suggestion in the penultimate paragraph that the fossil hunters may soon provide us with additional evidence of an earlier form that is more like man than the "australopithecines, including so-called *Homo habilis* and *Homo africanus*," a number of just such finds seem to have been made. Thus, to the skull fragment of Leakey, the talus from East Rudolf, and the humeral fragment from Kanapoi described above can now be added fragments of human jawbones and teeth from the Awash Valley in Ethiopia that date back three and four million years. These finds apparently comprise a complete upper jaw with all the teeth in place; a half upper jaw and a half mandible, both with teeth; and a variety of postcranial fragments. These finds have been made by Karl Johanson and Maurice Taieb, and they remind us that documenting the story of human evolution is a procedure that must continually change as new fossils are discovered. These fragments, older than most australopithecines, are described at least initially as more like man than australopithecines. They may therefore be thought of as "pre-man" and thus by their discovery may confirm the relegation, provided by the current study, of "*Homo habilis*" "*Homo africanus*," and other australopithecines to the status of "pseudo-man."

It is especially fascinating to me that investigative methods like those described in this book have proved capable of forecasting this situation. They are thus to be reckoned as a set of very powerful tools that have been added to the armamentarium of the evolutionary morphologist. Findings resulting from the application of such methods, once they have been thoroughly checked, should not lightly be discarded, even if they are at some variance with the conventional wisdom. Indeed such controversial results may well be the source of more thoughtful assessments of human evolution and may help us to escape from the bands that traditional thinking sometimes lays upon us.

References

Andrews, D. F. 1972. Plots of high-dimensional data. *Biometrics*, 28:125–36.

———. 1973. Graphical techniques for high dimensional data. In *Discriminant analysis and applications*. Ed. T. Cacoullos, pp. 37–59. New York: Academic Press.

Ashton, E. H.; Flinn, R. M.; Oxnard, C. E.; and Spence, T. F. 1975. The taxonomic and functional significance of overall bodily proportions of primates. *J. Zool.* 175.

Ashton, E. H.; Healy, M. J. R.; and Lipton, S. 1957. The descriptive use of discriminant functions in physical anthropology. *Proc. Roy. Soc. B.* 146:552–72.

Ashton, E. H.; Healy, M. J. R.; Oxnard, C. E.; and Spence, T. F. 1965. The combination of locomotor features of the primate shoulder girdle by canonical analysis. *J. Zool. Lond.* 147:406–29.

Ashton, E. H., and Oxnard, C. E. 1963. The musculature of the primate shoulder. *Trans. Zool. Soc. Lond.* 29:553–650.

———. 1964. Locomotor patterns in primates. *Proc. Zool. Soc. Lond.* 142:1–28.

Basmajian, J. V. 1972. Biomechanics of human posture and locomotion: Perspectives from electromyography. In *Functional and evolutionary biology of primates: Methods of study and recent advances*. Ed. R. H. Tuttle, pp. 292–304. Chicago: Aldine-Atherton.

Bean, R. B. 1922. The sitting height. *Amer. J. Phys. Anthrop.* 5:349–90.

Biegert, J., and Maurer, R. 1972. Rumpfskelettlänge, Allometrien und Körperproportionen bei catarrhinen Primaten. *Folia primat.* 17:142–56.

Bilsborough, A. 1971. Evolutionary change in the hominoid maxilla. *Man* 6:473–85.

Blackith, R. E., and Reyment, R. A. 1971. *Multivariate Morphometrics*. London: Academic Press.

Broom, R.; Robinson, J. T.; and Schepers, G. W. H. 1950. Sterkfontein ape-man *Plesianthropus*. *Transv. Mus. Mem.* 4:1–117.

Clark, Le Gros W. E. 1964. *The Fossil Evidence for Human Evolution*. 2d ed. Chicago: Univ. of Chicago Press.

———. 1967. *Man-Apes or Ape-Men*. New York: Holt, Rinehart and Winston.

Day, M. H. 1967. Olduvai hominid 10: a multivariate analysis. *Nature*, 215:323–24.

———. 1974. The interpolation of isolated fossil foot bones into a discriminant analysis—a reply. *Amer. J. Phys. Anthrop.* 41:233–36.

Day, M. H., and Napier, J. R. 1964. Hominid fossils from Bed I, Olduvai Gorge, Tanganyika: Fossil foot bones. *Nature*, 201:967–70.

Day, M. H., and Wood, B. A. 1968. Functional affinities of the Olduvai hominid 8 talus. *Man* 3:440–45.

———. 1969. Hominoid tali from East Africa. *Nature* 222:591–92.

Elliot, D. G. 1913. A review of the Primates. *Monogr. Amer. Mus. nat. Hist.* [3 vols.] 1:1–317, 2:1–382, 3:1–262.

Erikson, G. E. 1963. Brachiation in New World monkeys and in anthropoid apes. *Symp. Zool. Soc. Lond.* 10:135–64.

Farah, J. W.; Craig, R. G.; and Sikarskie, D. L. 1973. Photoelastic and finite element stress analysis of a restored axisymmetric first molar. *J. Biomech.* 6:511–20.

Frey, H. 1923. Untersuchungen über die Scapula, speziell über ihre äussere Form und deren Abhangigkeit von der Function. *Z. Anat. EntwGesch.* 68:277–324.

Goodman, M. 1973. The chronicle of primate phylogeny contained in proteins. *Symp. Zool. Soc. Lond.* 33:339–75.

Hall-Craggs, E. C. B. 1965. An analysis of the jump of the Lesser Galago (*Galago senegalensis*). *J. Zool.* 147:20–29.

Howells, W. W. 1972. Analysis of patterns of variation in crania of recent man. In *Functional and evolutionary biology of primates: Methods of study and recent advances.* Ed. R. H. Tuttle, pp. 123–51. Chicago: Aldine-Atherton.

————. 1973a. Cranial variation in man. (A study of multivariate analysis of patterns of difference among recent human populations.) *Papers of the Peabody Mus. Archaeology and Ethnology*, 67:1–259. Cambridge: Harvard University.

————. 1973b. *Evolution of the Genus Homo.* Reading, Massachusetts: Addison-Wesley.

Inman, V. T.; Saunders, J. B. De C. M.; and Abbott, LeR. C. 1944. Observations on the function of the shoulder joint. *J. Bone and Jt. Surg.* 26:1–30.

Jenkins, F. A., Jr. 1972. Chimpanzee bipedalism: Cineradiographic analysis and implications for the evolution of gait. *Science*, 178:877–79.

Jolly, A. 1972. *The Evolution of Primate Behavior.* New York: Macmillan.

Lanyon, L. E. 1971. Strain in sheep lumbar vertebrae recorded during life. *Acta orthop. Scandinav.* 42:102–12.

————. 1972. *In vivo* bone strain recorded from thoracic vertebrae of sheep. *J. Biomech.* 5:277–81.

————. 1973. Analysis of surface bone strain in the calcaneus of sheep during normal locomotion. Strain analysis of the calcaneus. *J. Biomech.* 6:41–49.

Leakey. R. E. F. 1973a. Further evidence of Lower Pleistocene hominids from East Rudolf, North Kenya, 1972. *Nature* 242:170–73.

————. 1973b Evidence for an advanced Plio-Pleistocene hominid from East Rudolf, Kenya. *Nature* 242:477–50.

Leakey, R. E. F., and Wood, B. A. 1973. New evidence of the genus *Homo* from East Rudolf, Kenya. II *Amer. J. Phys. Anthrop.*, 39:355–68.

Lestrel, P. E. 1973. Fourier analysis of cranial shape: a longitudinal study. [Abstract] *Proc. 42nd Ann. Mtg. Amer. Assn. Phys. Anthrop.*, Dallas, Texas. Ms in preparation.

Liem, K. F. 1970. Comparative functional anatomy of the Nandidae (Pisces: Teleostei). *Fieldiana Zool. Mem.* 56:1–166.

———. 1973. Evolutionary strategies and morphological innovations: Cichlid pharyngeal jaws. *Syst. Zool.* 22:425–41.

Lisowski, F. P. 1967. Angular growth changes and comparisons in the primate talus. *Folia primate.* 7:81–97.

Lisowski, F. P.; Albrecht, G. H.; and Oxnard, C. E. 1974a. A morphometric study of the talus in some higher primates, both living and fossil. [Abstract] *Amer. J. Phys. Anthrop.* 40:143.

———. 1974b. The form of the talus in some higher primates: a multivariate study. *Amer. J. Phys. Anthrop.* 41:191–215.

———. 1975. Further studies of African fossil tali. *Proc. Amer. Anthrop. Assoc.* In the press.

Lisowski, F, P., and Darlington, D. 1965. Certain dimensions of the primate talus. *Comm. to Int'l Anat. Congr.* Wiesbaden.

McHenry, H. M. 1972. The postcranial skeleton of early Pleistocene hominids. Ph.D. diss., Harvard University.

———. 1973. Early hominid humerus from East Rudolf, Kenya. *Science* 180: 739–41.

Mayr, E. 1963. *Animal species and evolution.* London: Oxford University Press.

Miller, R. A. 1932. Evolution of the pectoral girdle and forelimb in primates. *Amer. J. Phys. Anthrop.* 17:1–56.

Mollison, T. 1911. Die Körperproportionen der Primaten. *Morph. Jb.* 42:79–304.

Napier, J. R. 1959. Fossil metacarpals from Swartkrans. *Fossil Mammals of Afr.* no. 17:1–18. London: Brit. Mus. Nat. Hist.

Napier, J. R., and Napier, P. H. 1967. *A handbook of living primates.* London: Academic Press.

Nowinski, J. L., and Davis, C. F. 1970. A model of the human skull as a poroelastic spherical shell subjected to a quasistatic load. *Math. Biosci.* 8:397–416.

Oxnard, C. E. 1967. The functional morphology of the primate shoulder as revealed by comparative anatomical, osteometric and discriminant function techniques. *Amer. J. Phys. Anthrop.* 26:219–40.

———. 1968a. A note on the fragmentary Sterkfontein scapula. *Amer. J. Phys. Anthrop.* 28:213–17.

———. 1968b. A note on the Olduvai clavicular fragment. *Amer. J. Phys. Anthrop.* 29:429–31.

———. 1969a. Mathematics, shape and function: a study in primate anatomy. *Amer. Sci.* 57:75–96.

———. 1969b. Evolution of the human shoulder: some possible pathways. *Amer. J. Phys. Anthrop.* 30:319–31.

———. 1970. Functional morphology of primates: some mathematical and physical methods. *Burg-Wartenstein Symp.* No. 48:1–42.

Oxnard, C. E. 1972a. Functional morphology of primates: some mathematical and physical methods. In *Functional and evolutionary biology of primates: methods of study and recent advances*. Ed. R. H. Tuttle, pp. 305–36. Chicago: Aldine-Atherton.

———. 1972b. The use of optical data analysis in functional morphology: investigation of vertebral trabecular patterns. In *Functional and evolutionary biology of primates: methods of study and recent advances*. Ed. R. H. Tuttle, pp. 337–47. Chicago: Aldine-Atherton.

———. 1972c. Some African fossil foot bones: A note on the interpolation of fossils into a matrix of extant species. *Amer. J. Phys. Anthrop.* 37:3–12.

———. 1973a. *Form and Pattern in Human Evolution: Some mathematical, physical, and engineering methods*. Chicago: Univ. of Chicago Press.

———. 1973b. Some problems in the comparative assessment of skeletal forms. *Symp. Soc. Study Human Biol.* 11:103–25.

———. 1973c. Some locomotor adaptations among lower primates: implications for primate evolution. *Symp. Zool. Soc. Lond.* 33:255–99.

———. 1973d. Functional inferences from morphometrics: Problems posed by diversity and uniqueness among the primates. *Syst. Zool.* 22:409–24.

———. 1974. Primate structure and locomotion. *Amer. J. Phys. Anthrop.* 41:497–98.

———. 1975. Primate locomotor classifications for evaluating fossils: Their inutility, and an alternative. In *Proceedings from the Symposia of the Fifth Congress of the International Primatological Society*. Eds. S. Kondo; M. Kawai; A. Ehara; and S. Kawamura. Tokyo: Japan Science Press.

Patterson, B., and Howells, W. W. 1967. Hominid humeral fragment from early Pleistocene of northwestern Kenya. *Science* 156:64–66.

Pilbeam, D. 1972. *The ascent of man: An introduction to human evolution*. New York: Macmillan.

Preuschoft, H. 1972. Body posture and locomotion in some East African Miocene Dryopithecinae. *Symp. Soc. Study Human Biol.* 11:13–46.

Prieml. G. 1938. Die Platyrrhinen Affen als Bewegunstypen. Unter besonderer Berucksichtigung der Extremformen *Callicebus* und *Ateles. Z. Morph. Ökol. Tiere* 33:1–52.

Radinsky, L. 1967. Relative brain size: A new measure. *Science,* 155:836–38.

Rightmire, G. P. 1972. Multivariate analysis of an early hominid metacarpal from Swartkrans. *Science* 176:159–61.

Ripley. S. 1967. The leaping of langurs: a problem in the study of locomotor adaptation. *Amer. J. Phys. Anthrop.* 26:149–70.

Robinson, J. T. 1972. *Early Hominid Posture and Locomotion*. Chicago: Univ. of Chicago Press.

Rybicki, E. F.; Simonen, F. A.; and Weis, E. B., Jr. 1972. On the mathematical analysis of stress in the human femur. *J. Biomech.* 5:203–15.

Schultz, A. H. 1941. Growth and development of the orang-utan. *Contr. Embryol.* 29:57–110.

———. 1970. The comparative uniformity of the Cercopithecoidea. In *Old World monkeys: Evolution, systematics, and behavior.* Ed. J. R. Napier and P. H. Napier, pp. 39–51. London: Academic Press.

Shelman, C. B., and Hodges, D. 1970. A general purpose program for the extraction of physical features from a black and white picture. In *Symposium on feature extraction and selection in pattern recognition.* Ed. S. S. Yau and J. M. Garnett, pp. 135–44. New York: IEEE Computer Group.

Simons, E. L. 1972. *Primate evolution: An introduction to man's place in nature.* New York: Macmillan.

Sneath, P. H. A. 1967. Trend-surface analysis of transformation grids. *J. Zool. Lond.* 151:65–122.

Stern, J. T., Jr. 1971. Functional myology of the hip and thigh of cebid monkeys and its implications for the evolution of erect posture. *Bibliotheca Primatologica,* 14:1–318.

———. 1972. Anatomical and functional specializations of the human gluteus maximus. *Amer. J. Phys. Anthrop.* 36:315–39.

Stern. J. T., Jr., and Oxnard, C. E. 1973. Primate locomotion: some links with evolution and morphology. *Primatologia* 4(11):1–93.

Straus, W. L., Jr. 1948. The humerus of *Paranthropus robustus. Amer. J. Phys. Anthrop.* 6:285–311.

———. 1949. The riddle of man's ancestry. *Quart. Rev. Biol.* 24:200–223.

Thompson, D'Arcy W. 1917, 1942. *On Growth and form.* Cambridge: University Press.

Tuttle, R. H. 1969. Quantitative and functional studies on the hands of the Anthropoidea. I: The Hominoidea. *J. Morph.* 128:309–63.

———. 1972. Relative mass of cheiridial muscles in catarrhine primates. In *Functional and evolutionary biology of primates: methods of study and recent advances.* Ed. R. H. Tuttle, pp. 262–91. Chicago: Aldine-Atherton.

Uhlmann, K. 1968. Hüft-und Obserschenkelmuskulatur Systematische und vergleichende Anatomie. *Primatologia* 4(10):1–442.

Wanner, J. A. 1971. Relative brain size: A critique of a new measure. *Amer. J. Phys. Anthrop.* 35:255–57.

Wood, B. A. 1973. Locomotor affinities of hominoid tali from Kenya. *Nature* 246: 45–46.

———. 1974. Evidence on the locomotor pattern of *Homo* from early Pleistocene of Kenya. *Nature* 251:135–36.

Zuckerman, S.; Ashton, E. H.; Flinn, R. M.; Oxnard, C. E.; and Spence, T. F. 1973. Some locomotor features of the pelvic girdle in primates. *Symp. Zool. Soc. Lond.* 33:71–165.

Author Index

Species Index

Subject Index